Mathematics Online First Collections

This series covers the broad spectrum of theoretical, applied, and computational mathematics. Once peer-reviewed chapters are accepted for publication, they are published online ahead of the completion of the full volume. The readership for books in the series is intended to be made up of researchers and often times graduate students, as well. As in the publication, *Math in the Time of Corona*, the series may occasionally publish books fashioned for a general audience.

More information about this series at http://www.springer.com/series/16618

Tim Jax • Andreas Bartel • Matthias Ehrhardt •
Michael Günther • Gerd Steinebach

Editors

Rosenbrock—Wanner–Type Methods

Theory and Applications

 Springer

Editors
Tim Jax
Fachbereich Elektrotechnik. Maschinenbau
und Technikjournalismus
Hochschule Bonn-Rhein-Sieg
Sankt Augustin, Nordrhein-Westfalen
Germany

Andreas Bartel
Arbeitsgruppe Angewandte Mathematik
und Numerische Analysis
Bergische Universität Wuppertal
Wuppertal, Nordrhein-Westfalen, Germany

Matthias Ehrhardt
Arbeitsgruppe Angewandte Mathematik
und Numerische Analysis
Bergische Universität Wuppertal
Wuppertal, Nordrhein-Westfalen, Germany

Michael Günther
Arbeitsgruppe Angewandte Mathematik
und Numerische Analysis
Bergische Universität Wuppertal
Wuppertal, Nordrhein-Westfalen, Germany

Gerd Steinebach
Fachbereich Elektrotechnik. Maschinenbau
und Technikjournalismus und Institut für
Technik, Ressourcenschonung und
Energieeffizienz
Hochschule Bonn-Rhein-Sieg
Sankt Augustin, Nordrhein-Westfalen
Germany

ISSN 2730-633X   ISSN 2730-6348   (electronic)
Mathematics Online First Collections
ISBN 978-3-030-76809-6   ISBN 978-3-030-76810-2   (eBook)
https://doi.org/10.1007/978-3-030-76810-2

Mathematics Subject Classification: 65L04, 65L80, 65L05, 65L07, 65L20

This Springer imprint is published by the registered company Springer Nature Switzerland AG.
The registered company address is: Gewerbestrasse 11, 6330 Cham, Switzerland

Preface

In September 2017, the International Conference on Scientific Computation and Differential Equations (SCICADE) took place at the University of Bath, UK. Within this major meeting, the mini-symposium "Rosenbrock-Wanner-Type Methods: Theory and Applications" was organized by Tim Jax and Gerd Steinebach. It did attract 12 speakers underlining the research interest in this type of schemes. With that gathering of scientists, the idea for this book was born. That is, a monograph which summarizes current research and developments of Rosenbrock-Wanner (ROW) related methods.

Unfortunately, this project could not be realized immediately. With some delay, five still very relevant contributions are now available, which form this book.

Before we introduce the chapters, we recall that these methods are sometimes referred to as Kaps-Rentrop schemes. Moreover, if approximations of the Jacobians are incorporated directly into the design of these type of schemes, these techniques are called W-methods.

Chapter "Rosenbrock-Wanner Methods: Construction and Mission." Jens Lang in his chapter "Rosenbrock-Wanner Methods: Construction and Mission" gives a comprehensive overview of the development of the methods from the first ideas by Howard H. Rosenbrock in 1963 to current research activities. This chapter is suitable for newcomers in this field as well as for specialists. The reader gets an excellent introduction, but also a detailed insight into the variety of Rosenbrock-Wanner-derived methods. The complete bibliography provides an ideal starting point to further explore the world of Rosenbrock-Wanner-type methods.

Chapter "Water and Hydrogen Flow in Networks: Modelling and Numerical Solution by ROW Methods." In the chapter by Gerd Steinebach and David Michael Dreistadt, the focus is on the application of the methods. Infrastructure networks are considered, which are used for the transportation and storage of water and gas, and in particular hydrogen. These networks consist of individual components such as pipes, pumps, valves, metal-hydride tank, and so on. The mathematical modeling is based on conservation and reaction equations. In the case of spatially one-dimensional consideration, a semidiscretization in space is proposed, so that the modeling of all considered components leads to a large coupled DAE system. In two numerical case studies, the solution of the DAE systems by means of Rosenbrock-Wanner methods is discussed. Aspects such as smoothing of characteristic curves and switching processes, which play an important role in practice, are also discussed.

Chapter "Exponential Rosenbrock Methods and Their Application in Visual Computing." Vu Thai Luan and Dominik Michels introduce and apply explicit exponential Rosenbrock methods to large, (very) stiff systems of ordinary differential equations originating also from visual computing. Such systems can be coupled oscillators for elastodynamic systems. They investigate how these methods can be efficiently applied in this context. With numerical examples, they demonstrate the reduction of computational time when compared against state-of-the-art techniques.

Chapter "W-methods and Approximate Matrix Factorization for Parabolic PDEs with Mixed Derivative Terms." In the chapter by S. González-Pinto and D. Hernández-Abreu, W-methods are enriched with approximate matrix factorizations and three different families of methods are introduced. These methods are used for the time integration of parabolic partial differential equations, where mixed derivatives occur in the related elliptic operator. By the means of a scalar test equation, stability is investigated and demonstrated on numerical examples.

Chapter "Two-step W-methods." Two-step W-methods are investigated by Marcel Klinge, Helmut Podhaisky, and Rüdiger Weiner. These two-step schemes avoid lower-stage orders of classical ROW schemes and thus order reduction for stiff problems. In this chapter, the convergence theory is presented and order conditions as well as stability are discussed. A new family of schemes is constructed, which can be made stiffly accurate. Coefficient sets have been deduced for various convergence orders. These coefficient sets are numerically analyzed for different stiff test cases including Burgers' equation. The efficiency with respect to traditional Rosenbrock-Wanner schemes is demonstrated.

The editors would like to thank the publisher and authors of the individual chapters for their very good cooperation.

Sankt Augustin, Germany	Tim Jax
Wuppertal, Germany	Andreas Bartel
Wuppertal, Germany	Matthias Ehrhardt
Wuppertal, Germany	Michael Günther
Sankt Augustin, Germany	Gerd Steinebach
January 2021	

Contents

Rosenbrock-Wanner Methods: Construction and Mission

Jens Lang

1 Introduction

Howard Harry Rosenbrock (1920–2010) suggested in his famous paper from 1963 [55] to replace the iterative process for the solution of nonlinear problems within an implicit time integrator by a finite number of solutions of linear systems. He summarized: *Some general implicit processes are given for the solution of simultaneous first-order differential equations. These processes, which use successive substitution, are implicit analogues of the (explicit) Runge-Kutta processes. They require the solution in each time step of one or more sets of simultaneous linear equations, usually of a special and simple form. Processes of any required order can be devised, and they can be made to have a wide margin of stability when applied to a linear problem.* Thus, Rosenbrock methods avoid the problem of convergence for the solution of systems of nonlinear equations, making them a good alternative to fully implicit Runge-Kutta methods. In this note, I will give a brief historical overview and explain main construction principles including widely used members of the whole family of linearly implicit methods such as Rosenbrock-Wanner methods, W-methods, and recently developed two-step Rosenbrock-Peer and W-methods. Rosenbrock methods have made their way into real-life applications and become part of very sufficient adaptive multilevel PDE-solvers, see e.g. [27]. Nowadays, there still is an increasing interest in these methods, which would have delighted Rosenbrock who concluded his paper by expressing his wish: *The processes described above have been explored only cursorily, and it is hope that this note may stimulate others to investigate their possibilities.* It certainly did.

J. Lang (✉)
Technische Universität Darmstadt, Darmstadt, Germany
e-mail: lang@mathematik.tu-darmstadt.de

© The Author(s), under exclusive license to Springer Nature Switzerland AG 2021
T. Jax et al. (eds.), *Rosenbrock—Wanner–Type Methods*, Mathematics Online
First Collections, https://doi.org/10.1007/978-3-030-76810-2_1

Who was Howard H. Rosenbrock? Rosenbrock was born on December 16, 1920 in Ilford, England. He graduated 1941 from University College London with a 1st class honors degree in Electrical Engineering and received his PhD from London University in 1955. During the 1960s he worked at the Cambridge University and the MIT. In 1966, he became the Chair of Control Engineering at the University of Manchester, Institute of Science and Technology. He died on 21 October 2010. Rosenbrock produced over 120 scientific papers, 7 books, and about 30 papers on the philosophical basis of science and technology. An obituary was published in [80].

© IEEE Control System [80].

2 The Original Idea of Rosenbrock

In what follows, I will first review the original ideas of Rosenbrock as described in [55]. The writing is presented in a modern style, but a few text passages are included as pictures.

As starting point in his paper, Rosenbrock took a look at the (spatial) semi-discretization of the one-dimensional linear heat equation, i.e., formulas (1) and (2) in Fig. 1. He stated: *Any explicit numerical method of solving eqn. (2) (e.g. Runge-Kutta) replaces the exponentials by their truncated Taylor's series during one time interval of the solution. The exponentials tend to zero as t becomes large, whereas the truncated Taylor's series tend to infinity. A severe limitation on the length of the time intervals is thus introduced.*

solution of parabolic partial differential equations. If the equation is linear, for example

$$\frac{\partial x}{\partial t} = \kappa \frac{\partial^2 x}{\partial z^2} \tag{1}$$

and if it is replaced by a set of simultaneous ordinary differential equations

$$\dot{x} = \phi(x) = Ax \tag{2}$$

then the solution is a sum of terms containing exponentials $e^{-k_i t}$. Any explicit numerical method of solving

Fig. 1 © The Computer Journal, part of page 329 of [55]

To illustrate these fundamental observations, let us consider the heat equation in the form

$$
\begin{aligned}
\partial_t u &= \nabla \cdot (\mathbb{D} \nabla u), & x &\in \Omega & t &\in (0, T], \\
u(x, t) &= 0, & x &\in \partial\Omega, \ t \in (0, T], & & \\
u(x, 0) &= u_0(x), & x &\in \Omega, & &
\end{aligned}
\tag{1}
$$

with domain $\Omega \subset \mathbb{R}^d$ ($d \geq 1$) and a symmetric positive definite matrix $\mathbb{D}(x) \in \mathbb{R}^{d \times d}$. We have the following stability results:

$$
\|u(t)\|_{L^2(\Omega)} \leq \|u_0\|_{L^2(\Omega)}, \quad t \in [0, T],
\tag{2}
$$

$$
u(t) \to 0 \text{ for } t \to \infty.
\tag{3}
$$

A Method of Lines approach (let's take finite differences for simplicity) yields the system of ordinary differential equations

$$
\begin{aligned}
\partial_t U(t) &= A \, U(t), \ t \in (0, T], \\
U(0) &= U^0,
\end{aligned}
\tag{4}
$$

where the vector $U(t)$ collects approximations at certain spatial points. The matrix A is symmetric negative definite and therefore exhibits negative real eigenvalues— the values $-k_i$ in the exponentials mentioned by Rosenbrock in Fig. 1. An explicit Runge-Kutta methods computes approximations $U_n \approx U(t_n)$ with $t_n = nh$, $n \geq 1$ through

$$
\begin{aligned}
U_{n+1} &= R_{ERK}(hA) \, U_n, \quad n = 0, 1, \ldots \\
U_0 &= U^0,
\end{aligned}
\tag{5}
$$

where

$$
R_{ERK}(z) = 1 + z + \ldots + \frac{z^p}{p!} + \sum_{i=p+1}^{s} \alpha_i z^i = e^z + \mathcal{O}(z^{p+1}).
\tag{6}
$$

The stability requirements for the semi-discretized solution

$$
\|U_{n+1}\|_2 \leq \|U_n\|_2, \quad U_n \to 0 \text{ for } n \to \infty,
\tag{7}
$$

request $|R_{ERK}(z)| < 1$ for z along the negative real axis. Due to the nature of the approximation (6), small time steps h are necessary to guarantee stability. Moreover, the finer the spatial discretization, the smaller the time steps must be, showing that explicit methods, in general, are inefficient for the solution of such kind of (stiff) problems.

Crank and Nicholson (1947) pointed out that the restriction could be completely removed in linear problems by using the trapezoidal rule

$$k_r = h_r\{\tfrac{1}{2}\phi_{r-1} + \tfrac{1}{2}\phi_r\} \tag{3}$$

$$x_r' = x_{r-1}' + k_r \tag{4}$$

where h_r is the length of the rth time step, and the dash distinguishes the numerical from the exact solution of eqn. 2. This process replaces the exponentials occurring in the solution of eqn. (2) by the approximation

$$\psi_i(t) = \frac{1 - \tfrac{1}{2}k_i t}{1 + \tfrac{1}{2}k_i t} \tag{5}$$

during each time interval.

Fig. 2 © The Computer Journal, part of page 329 of [55]

An alternative is the implicit Crank-Nicolson method already proposed in 1947 [3], see also Fig. 2. It reads

$$U_{n+1} = R_{CN}(hA)\, U_n, \quad n = 0, 1, \ldots$$
$$U_0 = U^0, \tag{8}$$

with

$$R_{CN}(z) = \frac{1 + z/2}{1 - z/2} = e^z + \mathcal{O}(z^3). \tag{9}$$

The method is unconditionally stable, since $|R_{CN}(z)| \leq 1$ for all z lying in the left complex half plane. However, the damping properties at infinity are unsatisfactory. This lack has been also mentioned by Rosenbrock: *The procedure given in eqns. (3) and (4) has been widely used. It is perhaps not widely known, however, that instability can arise even with this process when the ϕ are non-linear functions of x. This is hardly surprising, since $\psi_i(t) \to -1$ as $t \to \infty$, so that even in a linear problem stability is only just maintained for large t.*

However, his main observation was that *when the functions ϕ are non-linear, implicit equations such as eqn. (3) can in general be solved only by iteration. This is a severe drawback, as it adds to the problem of stability, that of convergence of the iterative process.* As consequence, he set up a generalized implicit process with linear equations that can be solved rapidly and easily. How is it done?

Let us now consider the autonomous system of nonlinear ordinary differential equations

$$\partial_t U(t) = F(U(t)), \quad t \in (0, T],$$
$$U(0) = U^0.$$
(10)

and apply the Crank-Nicolson method to it, resulting in

$$U_{n+1} = U_n + \tfrac{h}{2} \left(F(U_n) + F(U_{n+1}) \right), \quad n = 0, 1, \dots$$
$$U_0 = U^0.$$
(11)

With U_n as starting values, Newton's method to approximate U_{n+1} gives the sequence of linear equations

$$U_{n+1}^{(0)} = U_n,$$
$$\left(I - \tfrac{h}{2} F'(U_{n+1}^{(k)}) \right) K_{n+1}^{(k+1)} = - \left(U_{n+1}^{(k)} - U_n - \tfrac{h}{2} \left(F(U_{n+1}^{(k)}) + F(U_n) \right) \right),$$
(12)
$$U_{n+1}^{(k+1)} = U_{n+1}^{(k)} + K_{n+1}^{(k+1)}, \quad k = 0, 1, \dots.$$

The fundamental idea of Rosenbrock was to use *only one step* of Newton's method, which reads for $k = 0$

$$\left(I - \frac{h}{2} F'(U_n) \right) K_{n+1} = h \, F(U_n),$$
(13)

or equivalently

$$K_{n+1} = h \left(F(U_n) + \tfrac{1}{2} F'(U_n) K_{n+1} \right).$$
(14)

In his paper, Rosenbrock did not mention how he derived his formula (9), see Fig. 3, and left it to the reader as an exercise.

process. An alternative to eqn. (3), which avoids this difficulty, is

$$(k_i)_r = h_r \left\{ (\phi_i)_{r-1} + \tfrac{1}{2} \sum_j \left(\frac{\partial \phi_i}{\partial x_j} \right)_{r-1} (k_j)_r \right\}.$$
(9)

This set of linear equations can be solved directly for k_r. and if each ϕ depends on only a few of the x the solution can be carried out rapidly and easily.

Fig. 3 © The Computer Journal, part of page 329 of [55]

A generalized implicit process may be obtained from eqn. (9) by analogy with Kopal's treatment of the Runge–Kutta processes (Kopal, 1955).   Let

$$A(x) = (A_{ij}(x)) = \left(\frac{\partial \phi_i(x)}{\partial x_j}\right) \tag{10}$$

and write

$$k_r = h_r\{\phi(x'_{r-1}) + a_1 A(x_{r-1})k\} \tag{11}$$

$$l_r = h_r\{\phi(x'_{r-1} + b_1 k_r) + a_2 A(x'_{r-1} + c_1 k_r)l_r\} \tag{12}$$

$$m_r = h_r\{\phi(x'_{r-1} + b_2 k_r + d_1 l_r) +$$
$$+ a_3 A(x'_{r-1} + c_2 k_r + e_1 l_r)m_r\} \tag{13}$$

$$\cdots$$

$$x'_r = x'_{r-1} + R_1 k_r + R_2 l_r + R_3 m_r + \cdots \tag{14}$$

Fig. 4 © The Computer Journal, part of page 329 of [55]

In a next step, he proposed to use Kopal's treatment of the Runge-Kutta processes [25] to design a generalized implicit process shown in Fig. 4. Note that the Jacobian is evaluated at different solutions. Rosenbrock did a consistency analysis: *By a straightforward but tedious calculation it is possible to expand $x'_r - x'_{r-1}$ in eqn. (14) as a power series in h_r, and to compare this with the Taylor's series.* He derived order conditions for two stages up to order four. Finally, I summarize his findings:

- There is no 2-stage third-order method with $R(\infty) = 0$.
- He constructed a 2-stage third-order method with $R(\infty) = -0.8$.
- He constructed a 2-stage second-order method with $R(\infty) = 0$.

Compared to the second-order Crank-Nicolson method, Rosenbrock found a 2-stage second-order method with optimal damping property at infinity and only two linear equations that have to be solved in each time step. General s-stage Rosenbrock or Rosenbrock-Runge-Kutta methods can be written in the (modern) form

$$\left(I - h\gamma_{ii} F'(U_n + \sum_{j=1}^{i-1} \delta_{ij} K_j)\right) K_i = h F(U_n + \sum_{j=1}^{i-1} \alpha_{ij} K_j), \quad i = 1, \ldots, s, \tag{15}$$

$$U_{n+1} = U_n + \sum_{i=1}^{s} b_i K_i.$$

This formulation is the starting point for further improvements.

3   The Improvement by Wanner

Around 1973, Gerhard Wanner became interested in Rosenbrock schemes and added his famous sum, $h F'(U_n) \sum_{j=1,\ldots,i-1} \gamma_{ij} K_j$, on the right hand side in (15), keeping at the same time the Jacobian fixed, i.e., using $F'(U_n)$ for all stages [77, 1977].

Rosenbrock-Wanner methods (short ROW methods) with s stages have the general form

$$\left(I - h\gamma_{ii} F'(U_n)\right) K_i = h F(U_n + \sum_{j=1}^{i-1} \alpha_{ij} K_j) + h F'(U_n) \sum_{j=1}^{i-1} \gamma_{ij} K_j,$$

$$i = 1, \ldots, s, \tag{16}$$

$$U_{n+1} = U_n + \sum_{i=1}^{s} b_i K_i.$$

In the spirit of Rosenbrock, they can be derived from diagonally implicit Runge-Kutta methods (short DIRK methods), applying only one simplified Newton step with the Jacobian $F'(U_n)$ and using already calculated stage values as starting values in the calculation of subsequent stages. Applied to (10), the nonlinear system for the stage values K_i of a DIRK method with lower triangular coefficient matrix $D = (d_{ij})$ reads

$$K_i = h F(U_n + \sum_{j=1}^{i} d_{ij} K_j), \quad i = 1, \ldots, s. \tag{17}$$

One-step of a Newton-like iteration

$$(I - h d_{ii} F'(U_n)) \left(K_i - K_i^{(0)}\right) = h F\left(U_n + \sum_{j=1}^{i-1} d_{ij} K_j + d_{ii} K_j^{(0)}\right) - K_i^{(0)} \tag{18}$$

with starting values

$$K_1^{(0)} = 0, \quad K_i^{(0)} = -\sum_{j=1}^{i-1} \frac{\gamma_{ij}}{d_{ii}} K_j, \quad i = 2, \ldots, s, \tag{19}$$

yields the ROW method (16) with $\alpha_{ij} = d_{ij} - \gamma_{ij}$ and $\gamma_{ii} = d_{ii}$. Compared to Rosenbrock's original form (15), the coefficients δ_{ij} were removed to avoid recalculations of Jacobians and new coefficients γ_{ij} were added to have enough parameters for consistency and good stability properties.

A usual simplification is to set $\gamma_{ii} = \gamma$ for all $i = 1, \ldots, s$. In case of direct solvers, it allows to reuse an LU-decomposition of the linear system matrix $I - h\gamma F'(U_n)$. It also simplifies iterative solvers, when matrix decompositions as preconditioners are used. To avoid the matrix-vector multiplication, one introduces $S_i = \sum_{j=1,\ldots,i} \gamma_{ij} K_j$ and solves

$$\left(\frac{I}{h\gamma} - F'(U_n)\right) S_i = F(U_n + \sum_{j=1}^{i-1} a_{ij} S_j) + \sum_{j=1}^{i-1} \frac{c_{ij}}{h} S_j,$$

$$i = 1, \ldots, s, \tag{20}$$

$$U_{n+1} = U_n + \sum_{i=1}^{s} m_i S_i.$$

Defining the matrix $\Gamma = (\gamma_{ij})_{i,j=1}^s$ with $\gamma_{ii} \neq 0$ for all i, the new parameters are derived from

$$(a_{ij})_{i,j=1}^s = (\alpha_{ij})_{i,j=1}^s \Gamma^{-1}, \quad (c_{ij})_{i,j=1}^s = \mathrm{diag}(\gamma_{11}^{-1}, \ldots, \gamma_{ss}^{-1}) - \Gamma^{-1},$$
$$(m_1, \ldots, m_s) = (b_1, \ldots, b_s) \Gamma^{-1}. \tag{21}$$

Further generalizations to non-autonomous systems and systems of the special multiplicative form $M(t, U)\partial_t U = F(t, U)$, where M might be singular, are also possible [15, 36].

So far, the Jacobian has to be computed at every time step, which can be quite costly. Steihaug and Wolfbrandt [67, 1979] developed so-called W-methods that avoid exact Jacobians, i.e., $F'(U_n) \approx T_n$ with arbitrary matrix T_n. The idea is to keep the Jacobian unchanged over several time steps while still ensuring stability. Less restrictive time lagged approximations of the form $T_n \approx F'(U_n) + \mathcal{O}(h)$ were proposed by Scholz and Verwer [61, 1983], see also Scholz [58, 59, 1978/79], and Kaps and Ostermann [21, 40, 1988/89]. Rahunanthan and Stanescu recently discussed high-order W-methods [48, 2010]. They have been also applied to optimal control problems in Lang and Verwer [31, 2013].

The linear equations in (20) can be successively solved. Order conditions were derived by applying the theory of Butcher series. They can be found in Wolfbrandt [83, 1977], Kaps [20, 1977], Nørsett and Wolfbrandt [39, 1979], and Kaps and Wanner [23, 1981]. Further details and many more information are given in the books of Van der Houven [72, 1976] and Hairer and Wanner [15, 1991].

For later use, I briefly recall the definition of a few fundamental stability concepts. Applied to the famous scalar Dahlquist's test equation $y' = \lambda y$, $y_0 = 1$ with $\lambda \in \mathbb{C}$, a ROW method (as any other Runge-Kutta method) gives $U_{n+1} = R(z)U_n$, where $z = \lambda h$. The function $R(z)$ is called the stability function of the method and the set $S = \{z \in \mathbb{C} : |R(z)| \leq 1\}$ defines its stability domain. The exact solution of the test equation is stable in the entire negative complex half plane $C^- = \{z : Re(z) \leq 0\}$, and it seems likely that a numerical method should preserve this stability property. Dahlquist [4, 1963] called a method *A-stable* if $C^- \subset S$. If in addition $\lim_{z \to -\infty} R(z) = 0$, the method is called *L-stable*—a property that was introduced by Ehle [7, 1969] and guarantees a fast damping for those z having very large negative real parts. A convenient way to ensure L-stability for ROW methods is to require $\alpha_{si} + \gamma_{si} = b_i$ for $i = 1, \ldots, s$, and $\sum_j \alpha_{sj} = 1$. Such methods are called *stiffly accurate*. A weaker concept was established by Widlund [82, 1967] who called a method *A(α)-stable* if the sector $S_\alpha = \{z : |\arg(-z)| \leq \alpha, z \neq 0\}$ is contained in its stability region.

There are A-stable and L-stable ROW methods available. ROW methods share their linear stability properties with (singly) diagonally implicit Runge-Kutta methods introduced by Alexander [1, 1977]. The role of the stability parameter γ was studied in Wanner [78, 1980]. Continuous extensions of Rosenbrock-type methods for a frequent graphical output were introduced by Ostermann [41, 1990].

4 Development of Rosenbrock-Wanner Methods

First Solvers The theoretical investigation of Rosenbrock-Wanner methods at the end of the 70s laid the starting point for a broad and fast development of efficient solvers. The fourth-order codes GRK4A and GRK4T proposed by Kaps and Rentrop [22, 1979] were equipped with a step size control based on embedded formulas of order three. The first one is A-stable whereas the second is only A(89.3^o)-stable, but comes with smaller truncation errors. They were successfully tested on the 25 stiff test problems of Enright et al. [8, 1975]. Gottwald and Wanner presented their back-stepping algorithm to improve the reliability of Rosenbrock methods [12, 1981]. Time-lagged Jacobian matrices and a modified Richardson extrapolation for variable steps size control within a fourth-order A-stable Rosenbock-Wanner scheme (named RKRMC) were tested by Verwer et al. [76, 1983]. Further analysis and experiments have been made by Verwer [74, 75, 1982]. Implementation issues were discussed by Shampine [63, 1982]. Veldhuizen investigated the D-stability of the Kaps-Rentrop methods [73, 1984]. There are two options to estimate local errors: embedding and Richardson extrapolation. Kaps, Poon, and Bui did a careful comparison of these two strategies in [24, 1985]. The performance of Rosenbrock methods for large scale combustion problems discretized by the Method of Lines was investigated by Ostermann et al. [42, 1986].

Partitioned Methods It is often useful to split the solution vector $U(t)$ into stiff and non-stiff components, say $U_s(t)$ and $U_n(t)$. After an appropriate reordering of the original equations, this gives a partitioned system

$$
\begin{aligned}
U_s'(t) &= F_s(U_s(t), U_n(t)), & U_s(0) &= U_s^0, \\
U_n'(t) &= F_n(U_s(t), U_n(t)), & U_n(0) &= U_n^0.
\end{aligned}
\tag{22}
$$

Now it is quite natural to apply a Rosenbrock-type scheme to the stiff part and an explicit Runge-Kutta method to the non-stiff part. Rentrop combined an A-stable Rosenbrock (3)4-pair with a common (4)5-Runge-Kutta-pair and studied strategies for stiffness detection in [52, 1985]. A drawback of such an approach is the occurrence of additional coupling conditions which usually does not allow the simple combination of two favourite schemes. An alternative is to use the setting of W-methods to directly incorporate the partitioning on the level of the Jacobian calculation, e.g., only take into account derivatives of F_s and drop the other ones. Such methods were analysed by Strehmel et al. [70, 1990] under the heading partitioned linearly implicit Runge-Kutta methods including ROW- and W-methods. Later on, Wensch designed an eight-stage fourth-order partitioned Rosenbrock method for multibody systems in index-3 formulation [81, 1998].

The partitioning can be also used to set up multirate schemes, where different step sizes for active and latent components are explicitly introduced in the discretization. In Günther and Rentrop [13, 1993], multirate Rosenbrock-Wanner methods were used for the simulation of electrical networks. One general shortcoming of multirate methods is the coupling between the components by interpolating and extrapolating state variables. Stability of multirate Rosenbrock methods were studied in Savcenco [56, 57, 2008/09] and Kuhn and Lang [26, 2014].

Differential-Algebraic Equations In the late 80s, Rosenbrock methods were also applied to differential-algebraic equations (DAEs) of index one:

$$\begin{aligned} U'(t) &= F(U(t), Z(t)), \quad U(0) = U^0, \\ 0 &= G(U(t), Z(t)), \quad Z(0) = Z^0, \end{aligned} \tag{23}$$

where it is assumed that $(\partial_Z G)^{-1}$ exists and is bounded in a neighbourhood of the solution. The main idea used by Roche [54, 1988] is to add $\varepsilon Z'(t)$ on the left hand side of the second equation and consider the DAE (23) as a limit case of the stiff singular perturbation problem for $\varepsilon \to 0$. This limit typically destroys the classical order of the Rosenbrock methods and gives rise to a new consistency theory derived by means of a modified Butcher-like tree model for the U- and Z-components. Note that the Kaps-Rentrop methods from [22] drop down to order two when applied to (23). Similar observations have been made earlier by Verwer [75, 1982]. Two new ROW-methods (named DAE34 and RKF4DA) with stepsize control and an index-1 monitor were proposed and tested by Rentrop et al. [53, 1989].

A desirable property when solving stiff or differential-algebraic equations is to have an L-stable method, i.e., a method with $R(\infty) = 0$. This is always the case for stiffly accurate Rosenbrock methods which approximate the algebraic component Z of the extreme DAEs, $U' = 1$ and $0 = G(U, Z)$, through one simplified Newton iteration. This nicely meets the original idea of Rosenbrock. In their book, Hairer and Wanner [15, 1991] constructed the famous stiffly accurate six-stage fourth-order Rosenbrock solver RODAS with an embedded method of order three. Special index-2 DAEs were treated in Lubich and Roche [36, 1990] and results for index-3 multibody systems can be found in Wensch [81, 1998]. Günther, Hoschek, and Rentrop constructed special index-2 Rosenbrock methods for electric circuit simulations [14, 2000]. Recently, Jax and Steinebach [18, 2017] introduced a new type of ROW methods for solving DAEs of the form (23). Taking ideas from W-methods, they allow arbitrary approximations to Jacobian entries resulting from the differential part.

Extrapolation An interesting, general approach to construct higher order methods for differential as well as differential-algebraic equations is to use extrapolation. Deuflhard and Nowak [6, 1987] proposed to extrapolate the linearly implicit Euler discretization (as the simplest Rosenbrock method) to solve chemical reaction kinetics and electric circuits and implemented the well-known variable-order LIMEX code with step size control. They also provided the impetus for Lubich to explain the error behaviour of such methods by perturbed asymptotic analysis [34, 1989].

B-Convergence and Order Reduction One-step methods and so Rosenbrock schemes suffer from order reduction, especially when they are applied to nonlinear parabolic partial differential equations. Sharp error estimates showing fractional orders of convergence for Rosenbrock and W-methods were first established by Lubich et al. [35, 43, 1993/95]. This phenomenon is related to the B-convergence of linearly implicit methods studied by Strehmel and Weiner [68, 1987]. Barriers for the order of B-convergence were given by Scholz [60, 1989]. In their book, Strehmel and Weiner [69, 1992] gave convergence results for spatial discretizations of semilinear parabolic equations with constant operator and a Lipschitz continuous non-linearity. However, the B-convergence technique does not give the sharp fractional temporal convergence rates. It is now much better understood than before why (lower) fractional orders occur. This reduction is not induced by lack of smoothness of the solution but rather by the presence of powers of the spatial differential operators in the local truncation error. Concerning W-methods, the order reduction is more severe compared with Rosenbrock methods. Loss of accuracy happens long before stability is affected. Fortunately, there are additional consistency conditions that imply also higher order of convergence as shown in Lubich and Ostermann [35, 1995].

Using this theoretical framework, new methods were constructed. Steinebach improved the RODAS code and designed his stiffly accurate RODASP scheme, which satisfies the new conditions for linear parabolic problems to reach order four. It was successfully applied to forecast transport in rivers, see Steinebach and Rentrop [65, 2001]. New order-three methods with three, ROS3P, and four stages, ROS3PL, were constructed in Lang and Verwer [29, 2001] and Lang and Teleaga [28, 2008], respectively. The latter one is stiffly accurate and therefore suitable for differential-algebraic equations. It also satisfies the condition of a W-method with $\mathcal{O}(h)$-disturbance of the Jacobian, which makes numerical differentiation for its entries less sensitive with respect to roundoff errors. A bunch of newly designed third-order Rosenbrock W-methods for partial differential-algebraic equations was published in Rang and Angermann [51, 2005]. Further improved ROW methods can be found in Rang [49, 50, 2014/15].

Exponential Rosenbrock-Type Methods Exponential integrators are based on a continuous linearization of the nonlinearity $F(U(t))$ along the numerical solution. This gives the linearized system

$$U'(t) = F'(U_n)U(t) + G_n(U(t)), \quad G_n(U(t)) = F(U(t)) - F'(U_n)U(t). \quad (24)$$

Exponential Rosenbrock methods make direct use of $J_n := F'(U_n)$ and $G_n(U(t))$. Hochbruck et al. [16, 2009] considered the following class of methods (here for variable time steps h_n):

$$U_{ni} = e^{c_i h_n J_n} U_n + h_n \sum_{j=1}^{i-1} a_{ij}(h_n J_n) g_n(U_{nj}), \quad i = 1, \ldots, s,$$

$$U_{n+1} = e^{h_n J_n} U_n + h_n \sum_{i=1}^{s} b_i(h_n J_n) g_n(U_{ni}). \quad (25)$$

A key point is the efficient approximation of the matrix exponential times a vector by Krylov subspace methods or methods based on direct polynomial interpolation.

An interpolation method with real Leja points was tested by Caliari and Ostermann [2, 2009] and showed a great potential for problems with large advection in combination with moderate diffusion and mildly stiff reactions. Higher order and parallel exponential Rosenbrock methods were proposed by Luan and Ostermann [32, 33, 2014/16].

Miscellaneous Rosenbrock methods offer a simple usage due to their linear structure. Methods up to order four perform well for low and medium tolerances and work competitive in many applications. The code ode23s in the MATLAB ODE SUITE is a typical Rosenbrock scheme, see Shampine and Reichelt [64, 1997]. The Krylov-W-code ROWMAP based on the Rosenbrock method ROS4 of Hairer and Wanner has demonstrated its efficiency for large stiff systems. Numerical tests were performed in Weiner et al. [79, 1997]. Rosenbrock methods are the numerical kernel in the adaptive multilevel PDAE-solver KARDOS, which is a well running working horse for a broad range of real-life applications, see Lang [27, 2000]. Combined with a linearized error transport equation based on first variational principles, they can be accompanied with a cheap global error estimation and control through tolerance proportionality. Such strategies were investigated in Lang and Verwer [30, 2007] for initial value problems and in Debrabant and Lang [5, 2015] for semilinear parabolic equations. Last but not least, a Rosenbrock code is listed in the second edition of *Numerical Recipes* by Press et al. [47, 1996].

A lot of basic information about Rosenbrock methods can be found in the books by Hairer and Wanner [15, 1991] and Strehmel and Weiner [69, 1992]. Newer developments are highlighted in Strehmel et al. [71, 2012]. A tremendous source of further interesting material are the proceedings of the numerous NUMDIFF-conferences held at the Martin Luther University Halle-Wittenberg since the early 1980s.

5   Two-Step Rosenbrock-Peer and W-Methods

As explained above, Rosenbrock methods may suffer from order reduction for very stiff problems. A closer inspection reveals that the low stage order (the first stage value is computed by the linearly implicit Euler scheme) is one of the reasons. To raise the stage order substantially, Podhaisky et al. [44, 45, 2002] studied a new class of linearly implicit two-step methods, where the previously computed stage values are taken into account. Such s-stage two-step W-methods have the form

$$Y_{ni} = U_n + h_n \sum_{j=1}^{s} a_{ij} U_{n-1,j} + h_n \sum_{j=1}^{i-1} \tilde{a}_{ij} U_{nj},$$

$$(I - \gamma h_n T_n) U_{ni} = F(Y_{ni}) + h_n T_n \sum_{j=1}^{s} \gamma_{ij} U_{n-1,j} + h_n T_n \sum_{j=1}^{i-1} \tilde{\gamma}_{ij} U_{nj},$$

$$i = 1, \ldots, s,$$

$$U_{n+1} = U_n + h_n \sum_{i=1}^{s} \left(b_i U_{ni} + v_i U_{n-1,i} \right).$$

$$(26)$$

Observe that $a_{ij} = \gamma_{ij} = v_i = 0$ recovers classical one-step ROW and W-methods. The special setting $\tilde{a}_{ij} = \tilde{\gamma}_{ij} = 0$ treated in [44] allows to compute the stage values U_{ni} in parallel. Higher order parallel methods were studied by Jackiewicz et al. [17, 2004]. Computer architectures of workgroup servers having shared memory for quite a few processors are particularly suitable for these methods which have been designed for the solution of large stiff systems in combination with Krylov techniques. Methods with favorable stability properties have been constructed with stage order $q = s$ and order $p = s$ for $s \leq 4$. All methods are competitive with state-of-the-art codes for stiff ODEs.

Within the class of two-step methods, Podhaisky et al. [46, 2005] also constructed s-stage methods where all stage values have the stage order $q = s - 1$. They considered the following methods:

$$(I - \gamma h_n T_n)U_{ni} = \sum_{j=1}^{s} b_{ij} U_{n-1,j} + h_n \sum_{j=1}^{s} a_{ij} \left(F(U_{n-1,j}) - T_n U_{n-1,j} \right)$$

$$+ h_n T_n \sum_{j=1}^{i-1} g_{ij} U_{nj}, \quad i = 1, \dots, s.$$

$$\tag{27}$$

Here, $U_{ns} \approx U(t_{n+1})$ and the matrix T_n is supposed to be an approximation to the Jacobian $F'(U(t_n))$ for stability reasons. The method is treated as a W-method, i.e., the order conditions are derived for arbitrary T_n. Due to their two-step and linear structure, the methods are called two-step Rosenbrock-Peer methods, where *peer* refers to the fact that all stage values have now one and same order. The methods constructed in [46] for $s = 4, \dots, 8$ are zero-stable for arbitrary step size sequences and L(α)-stable with large α. For constant time steps, these methods have order s. Numerical experiments showed no order reduction and an efficiency superior to the fourth-order RODAS for more stringent tolerances.

With this property, peer methods commend themselves as time-stepping schemes for the solution of time-dependent partial differential equations. So they have been implemented in the already mentioned finite element software package KARDOS, see Gerisch et al. [9, 2009] and Schröder et al. [62, 2017]. They also performed well for compressible Euler equations, demonstrated in Jebens et al. [19, 2012], for shallow-water equations, reported in Steinebach and Weiner [66, 2012], and for more complex fluid dynamics problems, see Gottermeier and Lang [10, 11, 2009/10]. More recently, linearly implicit two-step Peer methods of Rosenbrock-type have shown their reliability, robustness, and accuracy for large eddy and direct numerical simulations for turbulent unsteady flows in Massa et al. [37, 2018].

6 Summary

The idea of Rosenbrock is still alive. Avoiding the (often cumbersome) solution of nonlinear equations has not lost its attractiveness and significance over the years. The successive solution of linear equations is still a valuable option to efficiently solve systems of differential, differential-algebraic or partial differential equations. Classical one-step Rosenbrock-Wanner methods up to order four have demonstrated their good performance for low and medium tolerances. The new class of two-step Rosenbrock-Peer methods allows the construction of even higher order methods that overcome the disadvantage of order reduction and still exhibit good stability properties. Recent numerical experiments with higher tolerances are very promising.

There is still an ongoing research activity in the field of Rosenbrock methods. A recent search in the SCOPUS data base gave 753 documents. One of the last entries is about *Strong Convergence Analysis of the Stochastic Exponential Rosenbrock Scheme for the Finite Element Discretization of Semilinear SPDEs Driven by Multiplicative and Additive Noise* by Mukam and Tambue [38, 2018]. This brings me to my final remark. In view of the numerous contributions to Rosenbrock schemes, I would like to apologize in advance to those who have made significant further contributions to the topic but were not mentioned in my overview. I am prepared to receive your emails.

Acknowledgments I would like to thank Rüdiger Weiner for careful reading of a first version of the manuscript. His suggestions very much helped to improve the article. I also thank the reviewers for their useful remarks.

References

1. R. Alexander, Diagonally implicit Runge-Kutta methods for stiff ODE's. SIAM J. Numer. Anal. **14**, 1006–1021 (1977)
2. M. Caliari, A. Ostermann, Implementation of exponential Rosenbrock-type integrators. Appl. Numer. Math **59**, 568–581 (2009)
3. J. Crank, P. Nicolson, A practical method for numerical evaluation of solutions of partial differential equations of the heat-conduction type. Proc. Camb. Philos. Soc. **43**, 50–67 (1947)
4. G. Dahlquist, A special stability problem for linear multistep methods. BIT Math. Numer. **3**, 27–43 (1963)
5. K. Debrabant, J. Lang, On asymptotic global error estimation and control of finite difference solutions for semilinear parabolic equations. Comput. Methods Appl. Mech. Eng. **288**, 110–126 (2015)
6. P. Deuflhard, U. Nowak, Extrapolation integrators for quasilinear implicit ODE's, in *Large Scale Scientific Computing*, vol. 7, ed. by P. Deuflhard, B. Engquist. *Progress in Scientific Computing* (Birkhäuser, Basel, 1987), pp. 37–50
7. B.L. Ehle, On Padé approximations to the exponential function and A-stable methods for the numerical solution of initial value problems, CSRR 2010 (Dept. AACS, University of Waterloo, Ontario, Canada, 1969)
8. W. Enright, T.E. Hull, B. Lindberg. Comparing numerical methods for stiff systems of ODE's. BIT Numer. Math. **15**, 10–48 (1975)
9. A. Gerisch, J. Lang, H. Podhaisky, R. Weiner. High-order linearly implicit two-step peer— finite element methods for time-dependent PDEs. Appl. Numer. Math. **59**, 624–638 (2009)
10. B. Gottermeier, J. Lang, Adaptive two-step peer methods for incompressible Navier-Stokes equations, in *Numerical Mathematics and Advanced Applications 2009*, vol. 2, ed. by G. Kreiss, P. Lötstedt, A. Malqvist, M. Neytcheva. *Proceedings of ENUMATH 2009, Uppsala, Sweden* (2009), pp. 387–395

11. B. Gottermeier, J. Lang, Adaptive two-step peer methods for thermally coupled incompressible flow, in *Proceedings of the V European Conference on Computational Fluid Dynamics ECCOMAS CFD 2010, Lisbon, Portugal, 14–17 June 2010*, ed. by J.C.F. Pereira, A. Sequeira, J.M.C. Pereira (2010)
12. B.A. Gottwald, G. Wanner, A reliable Rosenbrock integrator for stiff differential systems. Computing **26**, 335–360 (1981)
13. M. Günther, P. Rentrop, Multirate ROW methods and latency of electric circuits. Appl. Numer. Math. **13**, 83–102 (1993)
14. M. Günther, M. Hoschek, P. Rentrop, Differential-algebraic equations in electric circuit simulation. Int. J. Electron. Commun. **54**, 101–107 (2000)
15. E. Hairer, G. Wanner, *Solving Ordinary Differential Equations II, Stiff and Differential-Algebraic Problems* (Springer, Berlin, 1991)
16. M. Hochbruck, A. Ostermann, J. Schweitzer, Exponential Rosenbrock-type methods. SIAM J. Numer. Anal. **47**, 786–803 (2009)
17. Z. Jackiewicz, H. Podhaisky, R. Weiner, Construction of highly stable two-step W-methods for ordinary differential equations. J. Comput. Appl. Math. **167**, 389–403 (2004)
18. T. Jax, G. Steinebach, Generalized ROW-type methods for solving semi-explicit DAEs of index-1. J. Comput. Appl. Math. **316**, 213–228 (2017)
19. S. Jebens, O. Knoth, R. Weiner, Linearly implicit peer methods for the compressible Euler equations. Appl. Numer. Math. **62**, 1380–1392 (2012)
20. P. Kaps, Modifizierte Rosenbrockmethoden der Ordnungen 4, 5 und 6 zur numerischen Integration steifer Differentialgleichungen. Ph.D. Thesis, Universität Innsbruck, Österreich, 1977
21. P. Kaps, A. Ostermann, Rosenbrock methods using few LU-decompositions. IMA J. Numer. Anal. **9**, 15–27 (1989)
22. P. Kaps, P. Rentrop, Generalized Runge Kutta methods of order four with step size control for stiff ODE's. Numer. Math. **33**, 55–68 (1979)
23. P. Kaps, G. Wanner, A study of Rosenbrock methods of high order. Numer. Math. **38**, 279–298 (1981)
24. P. Kaps, S.W.H. Poon, T.D. Bui, Rosenbrock methods for stiff ODEs: a comparison of Richardson extrapolation and embedding technique. Computing **34**, 17–40 (1985)
25. Z. Kopal, *Numerical Analysis* (Chapman and Hall, London, 1955)
26. K. Kuhn, J. Lang, Comparison of the asymptotic stability for multirate Rosenbrock methods. J. Comput. Appl. Math. **262**, 139–149 (2014)
27. J. Lang, *Adaptive Multilevel Solution of Nonlinear Parabolic PDE Systems. Theory, Algorithm, and Applications*. Lecture Notes in Computational Sciences and Engineering, vol. 16 (Springer, Berlin, 2000)
28. J. Lang, D. Teleaga, Towards a fully space-time adaptive FEM for magnetoquasistatics. IEEE Trans. Magn. **44**, 1238–1241 (2008)
29. J. Lang, J.G. Verwer, ROS3P—an accurate third-order Rosenbrock solver designed for parabolic problems. BIT Numer. Math. **41**, 730–737 (2001)
30. J. Lang, J.G. Verwer, On global error estimation and control for initial value problems. SIAM J. Sci. Comput. **29**, 1460–1475 (2007)
31. J. Lang, J.G. Verwer, W-methods in optimal control. Numer. Math. **124**, 337–360 (2013)
32. V.T. Luan, A. Ostermann, Exponential Rosenbrock methods of order five—construction, analysis and numerical comparisons. J. Comput. Appl. Math. **255**, 417–431 (2014)
33. V.T. Luan, A. Ostermann, Parallel exponential Rosenbrock methods. Comput. Math. Appl. **71**, 1137–1150 (2016)
34. Ch. Lubich, Linearly implicit extrapolation methods for differential-algebraic systems. Numer. Math. **55**, 197–211 (1989)
35. Ch. Lubich, A. Ostermann, Linearly implicit time discretization of non-linear parabolic equations. IMA J. Numer. Anal. **15**, 555–583 (1995)
36. Ch. Lubich, M. Roche, Rosenbrock methods for differential-algebraic systems with solution-dependent singular matrix multiplying the derivative. Computing **43**, 325–342 (1990)

37. F.C. Massa, G. Noventa, M. Lorini, F. Bassi, A. Ghidoni, High-order linearly implicit two-step peer methods for the discontinuous Galerkin solution of the incompressible Navier-Stokes equations. Comput. Fluids **162**, 55–71 (2018)
38. J.D. Mukam, A. Tambue, Strong convergence analysis of the stochastic exponential Rosenbrock scheme for the finite element discretization of semilinear SPDEs driven by multiplicative and additive noise. J. Sci. Comput. **74**, 937–978 (2018)
39. S.P. Nørsett, A. Wolfbrandt, Order conditions for Rosenbrock type methods. Numer. Math. **32**, 1–15 (1979)
40. A. Ostermann, Über die Wahl geeigneter Approximationen an die Jacobimatrix bei linear-impliziten Runge-Kutta Verfahren. Ph.D. Thesis, Universität Innsbruck, Österreich, 1988
41. A. Ostermann, Continuous extensions of Rosenbrock-type methods. Computing **44**, 59–68 (1990)
42. A. Ostermann, P. Kaps, T.D. Bui, The solution of a combustion problem with Rosenbrock methods. ACM Trans. Math. Software **12**, 354–361 (1986)
43. A. Ostermann, M. Roche, Rosenbrock methods for partial differential equations and fractional orders of convergence. SIAM J. Numer. Anal. **30**, 1084–1098 (1993)
44. H. Podhaisky, B.A. Schmitt, R. Weiner, Design, analysis and testing of some parallel two-step W-methods for stiff systems. Appl. Numer. Math. **42**, 381–395 (2002)
45. H. Podhaisky, R. Weiner, B.A. Schmitt, Two-step W-methods for stiff ODE systems. Vietnam J. Math. **30**, 591–603 (2002)
46. H. Podhaisky, R. Weiner, B.A. Schmitt, Rosenbrock-type 'Peer' two-step methods. Appl. Numer. Math. **53**, 409–420 (2005)
47. W.H. Press, S.A. Teukolsky, W.T. Vetterling, B.P. Flannery, *Numerical Recipes* (Cambridge University Press, Cambridge, 1996)
48. A. Rahunanthan, D. Stanescu, High-order W-methods. J. Comput. Appl. Math. **233**, 1798–1811 (2010)
49. J. Rang, An analysis of the Prothero-Robinson example for constructing new DIRK and ROW methods. J. Comput. Appl. Math. **262**, 105–114 (2014)
50. J. Rang, Improved traditional Rosenbrock-Wanner methods for stiff ODEs and DAEs. J. Comput. Appl. Math. **286**, 128–144 (2015)
51. J. Rang, L. Angermann, New Rosenbrock W-methods of order 3 for partial differential algebraic equations of index 1. BIT Numer. Math. **45**, 761–787 (2005)
52. P. Rentrop, Partitioned Runge-Kutta methods with stiffness detection and stepsize control. Numer. Math. **47**, 545–564 (1985)
53. P. Rentrop, M. Roche, G. Steinebach, The application of Rosenbrock-Wanner type methods with stepsize control in differential-algebraic equations. Numer. Math. **55**, 545–563 (1989)
54. M. Roche, Rosenbrock methods for differential algebraic equations. Numer. Math. **52**, 45–63 (1988)
55. H.H. Rosenbrock, Some general implicit processes for the numerical solution of differential equations. Comput. J. **5**, 329–330 (1963)
56. V. Savcenco, Comparison of the asymptotic stability properties for two multirate strategies. J. Comput. Appl. Math. **220**, 508–524 (2008)
57. V. Savcenco, Construction of a multirate RODAS method for stiff ODEs. J. Comput. Appl. Math. **225**, 323–337 (2009)
58. S. Scholz, S-stabile modifizierte Rosenbrock-Verfahren 3. und 4. Ordnung. Technical report, Sektion Mathematik, Technische Universität Dresden, 1978
59. S. Scholz, Modifizierte Rosenbrock-Verfahren mit genäherter Jacobi-Matrix. Technical report, Sektion Mathematik, Technische Universität Dresden, 1979
60. S. Scholz, Order barriers for the B-convergence of ROW methods. Computing **41**, 219–235 (1989)
61. S. Scholz, J.G. Verwer, Rosenbrock methods and time-lagged Jacobian matrices. Beiträge zur Numer. Math. **11**, 173–183 (1983)
62. D. Schröder, A. Gerisch, J. Lang, Space-time adaptive linearly implicit peer methods for parabolic problems. J. Comput. Appl. Math. **316**, 330–344 (2017)

63. L.F. Shampine, Implementation of Rosenbrock methods. ACM Trans. Math. Software **8**, 93–113 (1982)
64. L.F. Shampine, M.W. Reichelt, The MATLAB ODE Suite. SIAM J. Sci. Comput. **18**, 1–22 (1997)
65. G. Steinebach, P. Rentrop, An adaptive method of lines approach for modeling flow and transport in rivers, in *Adaptive Method of Lines*, ed. by A. Vande Wouwer, Ph. Saucez, W.E. Schiesser (Chapman Hall CRC, Boca Raton, 2001), pp. 181–205
66. G. Steinebach, R. Weiner, Peer methods for the one-dimensional shallow-water equations with CWENO space discretization. Appl. Numer. Math. **62**, 1567–1578 (2012)
67. T. Steihaug, A. Wolfbrandt, An attempt to avoid exact Jacobian and non-linear equations in the numerical solution of stiff differential equations. Math. Comput. **33**, 521–534 (1979)
68. K. Strehmel, R. Weiner, B-convergence results for linearly implicit one step methods. BIT Numer. Math. **27**, 264–281 (1987)
69. K. Strehmel, R. Weiner, *Linear-implizite Runge-Kutta-Methoden und ihre Anwendungen*. Teubner Texte zur Mathematik, vol. 127 (Teubner Stuttgart, Leipzig, 1992)
70. K. Strehmel, R. Weiner, I. Dannehl, On the behaviour of partitioned linearly implicit Runge-Kutta methods for stiff and differential algebraic systems. BIT Numer. Math. **30**, 358–375 (1990)
71. K. Strehmel, R. Weiner, H. Podhaisky, *Numerik gewöhnlicher Differentialgleichungen: Nichtsteife, steife und differential-algebraische Gleichungen*, 2nd edn. (Springer Spektrum, Springer, Berlin, 2012)
72. P.J. Van der Houwen, *Construction of Integration Formulas for Initial Value Problems* (North-Holland, Amsterdam, 1976)
73. M. Veldhuizen, D-stability and Kaps-Rentrop methods. Computing **32**, 229–237 (1984)
74. J.G. Verwer, An analysis of Rosenbrock methods for nonlinear stiff initial value problems. SIAM J. Numer. Anal. **19**, 155–170 (1982)
75. J.G. Verwer, Instructive experiments with some Runge-Kutta-Rosenbrock methods. Comput. Math. Appl. **7**, 217–229 (1982)
76. J.G. Verwer, S. Scholz, J.G. Blom, M. Louter-Nool, A class of Runge-Kutta-Rosenbrock methods for stiff differential equations. Z. Angew. Math. Mech. **63**, 13–20 (1983)
77. G. Wanner, On the integration of stiff differential equations, in *Numerical Analysis*, ed. by J. Descloux, J. Marti. *ISNM*, vol. 37 (Birkhäuser, Basel-Stuttgart, 1977), pp. 209–226
78. G. Wanner, On the choice of γ for singly implicit RK or Rosenbrock methods. BIT Math. Numer. **20**, 102–106 (1980)
79. R. Weiner, B.A. Schmitt, H. Podhaisky, ROWMAP—a ROW-code with Krylov techniques for large stiff ODEs. Appl. Numer. Math. **25**, 303–319 (1997)
80. P. Wellstead, Howard Harry Rosenbrock, Obituary. IEEE Control Syst. **31**, 89–101 (2011)
81. J. Wensch, An eight stage fourth order partitioned Rosenbrock method for multibody systems in index-3 formulation. Appl. Numer. Math. **27**, 171–183 (1998)
82. O.B. Widlund, A note on unconditionally stable linear multistep methods. BIT Math. Numer. **7**, 65–70 (1967)
83. A. Wolfbrandt, A study of Rosenbrock processes with respect to order conditions and stiff stability. Ph.D. Thesis, Chalmers University of Technology, Göteborg, 1977

Water and Hydrogen Flow in Networks: Modelling and Numerical Solution by ROW Methods

Gerd Steinebach and David Michael Dreistadt

1 Flow Problems in Networks

In this chapter fluid flow problems within networks are considered. The fluid can be water or gas, while the networks can be channels, rivers or pipes. Water flow in channels or rivers is characterized by a free surface, whereas in pipes free surface or pressurized flow may occur. Gas flow problems are restricted to pipe networks. In addition to the actual flow behaviour, the transport of substances and heat can also be modelled. Special attention is paid to a model of hydrogen storage that is based on metal hydride.

Applications with extensive experience by the authors include river water level forecast models [30], river alarm models for transport and dispersion of dangerous substances [22, 29], process simulation in sewer systems [31] and water supply network simulation [28, 32]. Further applications are e.g. gas flow through pipelines [14] and heat flow in district heating networks [7].

A common feature of these applications is the consideration of large networks where individual flow paths are modelled in one-dimension. These flow paths are coupled to the network nodes by suitable coupling conditions. The modelling is based on the conservation of mass, momentum and sometimes energy. This leads to hyperbolic conservation laws. If transport and dispersion processes of substances or

G. Steinebach (✉)
Hochschule Bonn-Rhein-Sieg, Institut für Technik, Ressourcenschonung und Energieeffizienz, Sankt Augustin, Germany
e-mail: gerd.steinebach@h-brs.de

D. M. Dreistadt
Helmholtz-Zentrum Geesthacht Zentrum für Material- und Küstenforschung GmbH, Geesthacht, Germany
e-mail: david.dreistadt@hzg.de

heat are to be considered, the conservation equations must be coupled with parabolic advection-diffusion equations via the flow velocities.

In addition to the actual flow paths represented by pipes, rivers or channels, other network elements can also be important (e.g. pumps, valves, storage tanks, heat exchangers). Such elements are modelled by ordinary differential equations (ODEs), or by algebraic equations. Together with the partial differential equations (PDEs) derived from the flow equations, a coupled system of partial differential algebraic equations (PDAEs) is given. Usually, the space derivatives of the PDEs are approximated by suitable finite differences or finite volumes. This approach leads to a large differential-algebraic equation (DAE) system, and consequently requires efficient numerical methods for their solution.

In the next section the modelling approach is discussed. Since water and gas flow problems are considered, a uniform approach for both is proposed. In Sect. 3 the numerical solution strategy is presented, based on the method of lines (MOL) with a conservative finite difference semidiscretization in space. For the solution of the resulting DAE-system, any efficient solver can, in principle, be applied. Because the models are often used in control rooms or forecasting centers, a reliable and robust scheme is required. In Sect. 4 numerical examples are considered and the results evaluated.

2   Unified Modelling Approach

To simplify the presented material, we will consider abstract networks that consist solely of nodes and directed edges, as exemplified in Fig. 1. Since water and gas flow is considered through the network, typical edges are pumps (or compressors), valves, control valves, connections and pipes. Specific to this example, a metal hydride hydrogen storage tank is also treated as an edge. Nodes can be simple

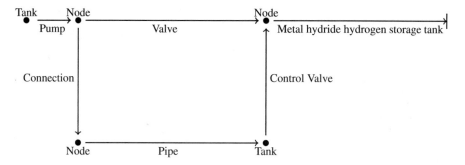

Fig. 1 An example water/gas network that consists of five nodes ("Tank" and "Node") and six edges ("Pump," "Valve," "Pipe," "Control Valve" and "Metal hydride hydrogen storage tank")

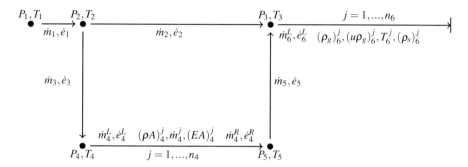

Fig. 2 Assignment of possible state variables to the network elements defined in Fig. 1. The lower index of the state variables represent the number of the corresponding node or edge. See text for variable definitions

coupling elements or tanks. A more detailed list of possible network elements can be found in reference [6] for gas networks and references [12, 33] for water networks.

After defining the element types, state variables and other characteristic values are assigned to the individual network elements. Characteristic values for nodes can be, for example, geographical coordinates or elevation data. For tanks, volume characteristics are required. Similarly, different characteristic values must be provided for the edges. Examples include pump characteristics, valve coefficients, pipe diameters, and so on. An illustration of state variable assignments to a network's elements is shown in Fig. 2. In order to enable uniform modelling of water and gas flows, the nodes are assigned the variables pressure P and temperature T. Moreover, to each node i, source terms can be assigned that represent given external mass flow $\dot{m}_{s,i}$ and energy flux $\dot{e}_{s,i}$. The quantities mass flow $\dot{m} = \rho \dot{q}$ with density ρ, volume flow $\dot{q}$ and energy flux $\dot{e}$ are assigned to the edges. Usually, a spatially constant mass and energy flux exists within an edge, see edges 1,2,3,5. For pipes, additional spatially resolved inner states are possible (see edge 4). In addition to the left and right side mass and energy flow, internal variables (ρA), $\dot{m} = \rho u A$ and (EA) with pipe flow cross-sectional area A, flow velocity u and total specific energy E must be present. Edge 6 represents a metal hydride hydrogen storage tank that can only be connected to a node from the left side. Thus, there is no need for right side states $\dot{m}_6^R$, $\dot{e}_6^R$. Within this storage tank the densities ρ_s of the solid phase and ρ_g of the gas phase must be distinguished. The total number of time-dependent unknowns within Fig. 2's example network is $N = 24 + 3 n_4 + 4 n_6$, where n_4 and n_6 are the numbers of the spatially resolved inner states of edges 4 and 6.

Up to now, the transport of chemical substance in the example network has not been explicitly considered. To do so, additional state variable c_i for the substance's concentration at each node i and mass flow $(c\dot{q})_j$ in each edge j could be introduced. Its transport through the pipe system is described by advection-diffusion equations.

2.1   Node Equations for State Variables

For the most part, nodes have no physical function. They are only used to link individual network elements preserving mass and energy conservation. Regarding the nodes 2,3 and 4 the corresponding coupling conditions read

$$0 = \dot{m}_1 - \dot{m}_2 - \dot{m}_3 + \dot{m}_{s,2} \tag{1}$$

$$0 = \dot{e}_1 - \dot{e}_2 - \dot{e}_3 + \dot{e}_{s,2} \tag{2}$$

$$0 = \dot{m}_2 + \dot{m}_5 - \dot{m}_6^L + \dot{m}_{s,3} \tag{3}$$

$$0 = \dot{e}_2 + \dot{e}_5 - \dot{e}_6^L + \dot{e}_{s,3} \tag{4}$$

$$0 = \dot{m}_3 - \dot{m}_4^L + \dot{m}_{s,4} \tag{5}$$

$$0 = \dot{e}_3 - \dot{e}_4^L + \dot{e}_{s,4} \tag{6}$$

Exceptions are nodes with a storage function (e.g. tanks). Assuming that nodes 1 and 5 are tanks, the following equations result:

$$\frac{d}{dt} M_1 = -\dot{m}_1 + \dot{m}_{s,1} \tag{7}$$

$$c_v \frac{d}{dt} (M_1 T_1) = -\dot{e}_1 + \dot{e}_{s,1} \tag{8}$$

$$\frac{d}{dt} M_5 = \dot{m}_4^R - \dot{m}_5 + \dot{m}_{s,5} \tag{9}$$

$$c_v \frac{d}{dt} (M_5 T_5) = \dot{e}_4^R - \dot{e}_5 + \dot{e}_{s,5} \tag{10}$$

To formulate these equations, it is assumed that the energy fluxes to a storage node are completely converted into heat. Here, c_v denote specific heat at constant fluid volume and M_1, M_5 are the fluid mass within the two storage tanks.

When the fluid is water and it is assumed that the tank is cylindric with base area A and a bottom elevation z over some reference level, the filling height h of the water is given by the hydrostatic pressure law $h = \frac{P}{\rho g}$ and the piezometric pressure height is $H = z + h$. The ρ symbol denotes a given constant density of water and g is the gravitational constant. The total mass in the tank is then given by $M = \rho A h$.

When the fluid is a gas, the ideal gas law is assumed:

$$P = \rho R_s T \tag{11}$$

with known specific gas constant $R_s = c_p - c_v$, where c_p is the specific heat at constant pressure. The total mass is then given by $M = \frac{P}{R_s T} V$, where V is the specified tank volume. Natural gas can be considered by a real gas factor $z(P, T)$ that depends on pressure and temperature, and by the corrected gas law $P = \rho R_s T z(p, T)$ (see reference [6, 16]).

2.2 Edge Equations for State Variables

Once nodes are mathematically defined, the equations for the edges must be specified. Each edge connects two nodes. The node to which the edge points is referred to as the right (R) node and the other as the left (L) node. In case of edge 1 the left node variables are $P_L = P_1$, $T_L = T_1$ and the right ones are $P_R = P_2$, $T_R = T_2$. An exception is edge 6, which represents a metal hydride hydrogen storage tank and is connected to a node only on the left side. In the following, only a few edge types are treated as examples.

2.2.1 Connections

A connection does not have any physical meaning. Nevertheless, connections can be applied to define a desired constant pressure difference ΔP, which may be caused by components which are not directly taken into account in the model, see references [12, 33]. In this case, the edge is modelled by the two equations

$$P_R - P_L = \Delta P \ , \tag{12}$$

$$T_L = T_R \ . \tag{13}$$

Here temperature effects due to the pressure difference are neglected. Since the unknowns $\dot{m}$ and $\dot{e}$ of the edge are not explicitly present in Eqs. (12) and (13), the DAEs of all node and edge equations might be of index two. This can easily be seen when two tanks are combined by an edge of type connection.

Usually, the energy flux being transmitted by an edge, except in the case of pipes, is restricted as heat flow. When the edge has no influence on the system's energy, the energy flux $\dot{e}$ is known and Eq. (13) is replaced by

$$\dot{e} = c_v \frac{1}{2} \left(|\dot{m}|(T_L - T_R) + \dot{m}(T_L + T_R) \right) \ . \tag{14}$$

By this equation the heat flow between the connected left and right nodes has the same direction as the mass flow.

2.2.2 Pumps

Water pumps are used to increase pressure and generate flow. A pump can be characterized by a relationship between pumping head ΔH and volume flow $\dot{q}$, also denoted as Q. Typically, a quadratic function is applied [33]:

$$\Delta H = \alpha_0 - \alpha_2 Q^2 . \tag{15}$$

Since water density ρ is known, $Q = \frac{\dot{m}}{\rho}$ and the pumping head is given by:

$$\Delta H = z_R + \frac{P_R}{\rho g} - z_L - \frac{P_L}{\rho g} .$$

The pumping process is assumed to be adiabatic and energy flux is given by Eq. (14).

For a gas, a compressor is similar to a water pump. Consequently, it can be characterized by a relationship between pressure difference and volume flow. In contrast to Eq. (15), this relation and the outflow temperature also depend on the inflow temperature:

$$Q = f_1(P_L, P_R, T_L) \quad \text{with} \quad Q = \frac{\dot{m}}{\rho_L} , \tag{16}$$

$$T_R = f_2(P_L, P_R, T_L) . \tag{17}$$

Inflow density ρ_L can again be computed by the ideal gas law from P_L and T_L. For details on the choice of functions f_1 and f_2 see [6, 16].

2.2.3 Valves

A valve leads to another type of pressure loss and Eq. (15) is replaced by:

$$s^2(t)(H_L(t) - H_R(t)) = \zeta |Q| Q . \tag{18}$$

Here, parameter ζ is a given pressure loss coefficient and $s(t) \in [0, 1]$ is the opening degree of the valve, which may vary with time (see reference [12, 33]). Equation (18) can be applied for both water and gas modelling. Usually, effects on temperature caused by the valve are neglected leading to Eq. (14) for energy flux.

2.2.4 Pipes

Pipes are an edge type that require careful modelling. Edge 4 in Fig. 2 represents a pipe. *Pipe flow of gas* is modelled in one-dimension space $x \in [x_L, x_R]$ with conservation of mass, momentum and energy leading to the Euler equations (see references [2, 10, 17]):

$$\frac{\partial(\rho A)}{\partial t} + \frac{\partial(u\rho A)}{\partial x} = 0 \,, \tag{19}$$

$$\frac{\partial(u\rho A)}{\partial t} + \frac{\partial(u^2\rho A)}{\partial x} + A\frac{\partial P}{\partial x} = -g\rho A \sin(\phi) - 2f\frac{\rho A}{D}|u|u \,, \tag{20}$$

$$\frac{\partial(AE)}{\partial t} + \frac{\partial}{\partial x}(uA(E+P)) = \Omega \,. \tag{21}$$

$A(x)$ denotes the known pipe cross-section with diameter $D(x)$, which may vary in space, and $\dot{m} = u\rho A$ is the mass flow with velocity u. The source terms on the right-hand side of the Euler equations describe the slope of the pipe with angle ϕ, wall friction with friction coefficient f and power losses or inputs Ω per unit of length.

The total specific energy E consists of internal, kinetic and potential energy. It is assumed that internal energy is proportional to temperature, leading to [17]:

$$E = \rho(c_v T + \frac{1}{2}u^2 + gz) \,. \tag{22}$$

In order to close the system, pressure P must be expressed as a function of the state variables. From Eq. (11) it follows that $T = \frac{P}{\rho(c_p - c_v)}$ and inserting into Eq. (22) yields

$$E = \frac{P}{\gamma - 1} + \frac{1}{2}\rho u^2 + \rho gz \quad \text{with } \gamma = \frac{c_p}{c_v}. \tag{23}$$

By discretizing the space interval $[x_L, x_R]$ into $x_L < x_1 < \ldots < x_n < x_R$, the chosen inner state variables $(\rho A)_i$, $\dot{m}_i$, $(AE)_i$, $i = 1, \ldots, n$ become obvious. Note, that the superscripts used in Fig. 2 appear as subscripts now. In order to later approximate the space derivatives by finite differences, these state variables must also be given on the boundaries x_L and x_R. The $\dot{m}_L$, $\dot{m}_R$ variables are defined separately and due to the neighboring nodes, T_L, P_L, T_R, P_R are given. Due to the ideal gas law, one gets ρ_L, ρ_R and the remaining state variables $(\rho A)_L$, $(AE)_L$, $(\rho A)_R$, $(AE)_R$ can be computed. Finally, $\dot{e}_L = c_v\dot{m}_L T_L$ and $\dot{e}_R = c_v\dot{m}_R T_R$ are defined.

In order to define $\dot{m}_L$ and $\dot{m}_R$, a numerical boundary condition must be applied to each side of the interval $[x_L, x_R]$. These boundary conditions can be derived from the invariants of the Euler equations. When neglecting the source terms on the right-hand side and assuming $A(x) = A$ to be constant, two invariants are given by [10]:

$$I_L(x(t), t) = u + \frac{2c}{\gamma - 1} = const \quad \text{for} \quad \frac{d}{dt}x(t) = u - c \,, \tag{24}$$

$$I_R(x(t), t) = u - \frac{2c}{\gamma - 1} = const \quad \text{for} \quad \frac{d}{dt}x(t) = u + c \,; \tag{25}$$

where $c = \sqrt{\frac{\gamma P}{\rho}}$ denotes the sound velocity. How these invariants are used for implementing the numerical boundary conditions is described in the next section.

When *free surface water flow* with constant density ρ and hydrostatic pressure law $P = \rho g h$ with water depth h is considered, Eqs. (19) and (20) are simplified to:

$$\frac{\partial A}{\partial t} + \frac{\partial Q}{\partial x} = 0, \tag{26}$$

$$\frac{\partial Q}{\partial t} + \frac{\partial}{\partial x}\left(\frac{Q^2}{A}\right) + gA\frac{\partial h}{\partial x} = -gA\sin(\phi) - 2f\frac{1}{AD}|Q|Q. \tag{27}$$

These equations are well known as the Saint-Venant equations [34]. It is assumed that there is a unique relationship between the flowed cross-sectional area $A(x, t)$ and the water depth $h(x, t)$. If the slope is small, the approximation $\sin(\phi) = \frac{d}{dx}z$ with a bottom elevation $z(x)$ can be used, and the terms $gA\frac{\partial h}{\partial x} + gA\frac{d}{dx}z = gA\frac{\partial H}{\partial x}$ are summarized with $H = z + h$ being the water surface level. The frequently used Manning-Strickler friction formula with coefficient K_{St} is obtained, when factor f is chosen as $f = \frac{gD}{2K_{St}^2 h^{4/3}}$. Moreover in case of water, only energy flux due to heat is considered. When $E = \rho c_v T$ is assumed, and source terms and temperature effects due to the pressure difference are neglected, Eq. (21) simplifies to the transport equation

$$\frac{\partial(AT)}{\partial t} + \frac{\partial(uAT)}{\partial x} = 0. \tag{28}$$

For *pressure flow of water* in pipes a Boussinesq assumption $dP = c^2 d\rho$ with sound velocity c of water is applied [3]. When the piezometric pressure height $H = z + D + \frac{P}{g\rho_0}$ with constant water density ρ_0 is introduced and a constant pipe diameter D is considered, the following relationships hold:

$$\frac{\partial H}{\partial t} = \frac{c^2}{g\rho_0}\frac{\partial \rho}{\partial t}, \quad \frac{\partial P}{\partial x} = g\rho_0\frac{\partial(H - z)}{\partial x}.$$

Dividing Eqs. (19) and (20) by constant A, neglecting the momentum term $\frac{\partial(u^2\rho A)}{\partial x}$, using $\sin(\phi) = \frac{d}{dx}z$ and dividing again by ρ leads to the well known water-hammer equations [1]:

$$\frac{\partial H}{\partial t} + \frac{c^2}{gA}\frac{\partial Q}{\partial x} = 0, \tag{29}$$

$$\frac{\partial Q}{\partial t} + gA\frac{\partial H}{\partial x} = -\lambda(Q)\frac{Q|Q|}{2DA}. \tag{30}$$

The friction term in this case is of type Darcy-Weisbach, [1]. As an example, this system of equations is applied in numerical tests in Sect. 4. The invariants (24) and

(25) now read as

$$I_L(x(t), t) = g A H - c Q = const \quad \text{for} \quad \frac{d}{dt} x(t) = -c, \tag{31}$$

$$I_R(x(t), t) = g A H + c Q = const \quad \text{for} \quad \frac{d}{dt} x(t) = c. \tag{32}$$

2.2.5 Special Case: Hydrogen Storage in Metal Hydride

With regards to using hydrogen as an energy carrier in gas networks, gas storage facilities are important components. In addition to a physically bound storage form, such as high-pressure gas or cryogenic storage, storage in a chemically bound form in metal hydride tanks is suitable. In metal hydride storage tanks, a reversible reaction occurs between the hydrogen gas and the metal, and thus binding the gas chemically. This process depends on pressure and temperature. In order to integrate hydrogen gas storage into a gas network model, the hydrogen flow is mathematically described by one-dimensional conservation equations that take into account the reaction kinetics. Mass conservation the hydrogen gas is described by Eq. (19), whereby the metal hydride storage tank is defined as a one-sided closed pipe filled with porous medium.

Introducing the porosity ϵ of the medium, gas density ρ_g and gas velocity u_g in pores, Eq. (19) becomes:

$$\frac{\partial (\epsilon A \rho_g)}{\partial t} + \frac{\partial (u_g \epsilon A \rho_g)}{\partial x} = -\dot{m}$$

with mass flow $\dot{m}$ per unit length generated by the storage of the gas. Since gas velocity in pores u_g is difficult to calculate, the Darcy velocity u is used to describe the gas flow and the connection between these states is the Dupuit-Forchheimer assumption [5]

$$u_g = \frac{u}{\epsilon}.$$

Assuming the pipe cross-section A and porosity ϵ to be constant, the mass conservation of the gas is given by

$$\epsilon \frac{\partial (\rho_g)}{\partial t} + \frac{\partial (u \rho_g)}{\partial x} = -\dot{\tilde{m}} \tag{33}$$

with specific mass flow $\dot{\tilde{m}} = \frac{1}{A} \dot{m}$. Since the hydride storage is assumed to be a porous medium, the law of Darcy replaces the momentum Eq. (20):

$$u = -\frac{K}{\mu} \frac{\partial P}{\partial x} \tag{34}$$

where K is the permeability of the porous material and μ the dynamic viscosity of the gas. Pressure P can be calculated by applying the ideal gas law (Eq. (11)).

In the mass balance of the solid phase with density ρ_s, convection does not occur:

$$(1 - \epsilon)\frac{\partial \rho_s}{\partial t} = \dot{\bar{m}} . \tag{35}$$

Due to the temperature dependence of the reaction kinetics, conservation of energy must be taken into account. In contrast to Euler equations, nearly constant pressure conditions are assumed and only heat energy is regarded. Thus, the total specific energy is given by

$$E = \rho c_p T$$

where c_p is the heat capacity at constant pressure. Moreover, gas and solid phases are assumed to have equal temperature denoted by T. This leads to the energy conservation (see reference [18]):

$$(\rho c_p)_e \frac{\partial T}{\partial t} + \rho_g c_{pg} u \frac{\partial T}{\partial x} = \lambda_e \frac{\partial^2 T}{\partial x^2} + \dot{\bar{m}}\left(\frac{\Delta H}{M_g} + T(c_{pg} - c_{ps})\right) + \dot{Q} \tag{36}$$

with averaged coefficients

$$(\rho c_p)_e = \epsilon \rho_g c_{pg} + (1 - \epsilon)\rho_s c_{ps} \quad \text{and} \quad \lambda_e = \epsilon \lambda_g + (1 - \epsilon)\lambda_s .$$

Here, thermal conductivities λ_g and λ_s of the gas and solid phases, reaction enthalpy ΔH and energy flow $\dot{Q}$ to the environment are taken into account. When a heat exchanger is used, energy flow can be described as a function of the heat transfer coefficient α:

$$\dot{Q} = \alpha A(T - T_a) , \tag{37}$$

where α depends, among other things, on the design of the heat exchanger and the design of the hydride storage tank and T_a denotes temperature of the heat transfer medium.

The specific mass flow $\dot{\bar{m}}$ depends on the reaction kinetics. Since only the macroscopic observation of the reaction is of interest in terms of the network simulation, it makes sense to use empirical models that are parameterized on the basis of measurement data. In a simple form, the specific mass flow that results from the reaction can be described by the following equations [11]:

$$\dot{m}_{abs} = C_a \exp\left(-\frac{E_a}{RT}\right) \ln\left(\frac{P}{P_{eq}}\right)(\rho_{ss} - \rho_s) , \tag{38}$$

$$\dot{m}_{des} = C_d \exp\left(-\frac{E_d}{RT}\right)\frac{P - P_{eq}}{P_{eq}}(\rho_s - \rho_{su}) , \tag{39}$$

where $\dot{m}_{abs}$ and $\dot{m}_{des}$ are the specific mass flow when absorbing or desorbing hydrogen. C_a and C_d are experimentally measured reaction rates, E_a and E_d are experimentally determined activation energies, R is the universal gas constant, ρ_{ss} is the saturated density and ρ_{su} the unsaturated density of the solid phase. For an ideal and reversible hydride formation process, the equilibrium pressure P_{eq} is assumed to be identical for both absorption and desorption. It can be compared to the gas pressure P for deciding if an absorption or desorption process takes place. In conclusion, the mass flow $\dot{m}$ can be calculated using the following equation:

$$\dot{m} = \frac{1}{2} \, \mathrm{sgn}(P - P_{eq})(\dot{m}_{abs} - \dot{m}_{des}) + \frac{1}{2}(\dot{m}_{abs} + \dot{m}_{des}) \,. \tag{40}$$

To calculate equilibrium pressure P_{eq} for an ideal and reversible hydride formation process, a correlation between the equilibrium pressure and the temperature must be defined. Figure 3 shows this relationship and the phases of ideal metal hydride formation [35].

In the first phase (α), hydrogen gas dissolves into the metal lattice under high increase of pressure, until a saturated concentration is reached. In the second phase (denoted as $\alpha + \beta$) the hydride formation takes place. Within this phase, the hydrogen concentration w in the solid phase increases at constant pressure (i.e. the equilibrium pressure P_{eq}) until the entire metal is converted into metal hydride. Notice that as P_{eq} increases, the length of the corresponding hydrogen-formation plateau decreases by increasing temperature until a critical temperature T_{kr} is reached. The reaction heat generated during hydride formation usually is dissipated so that the reaction does not come to a standstill. After complete absorption, only metal hydride is present (β phase). The desorption process is the reverse of the absorption process. It should also be noted that normally the pressure-plateaus

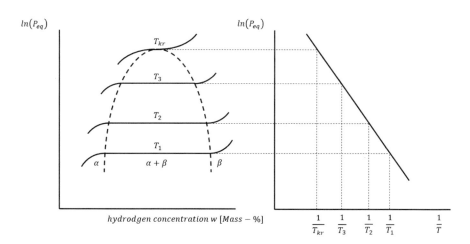

Fig. 3 Phases of hydride formation [35]

in Fig. 3 show an incline and hysteresis between the course of absorption and desorption. The van't Hoff equation can be derived from the equilibrium pressures that are present when the plateau is reached [4]:

$$\ln(P_{eq}) = \frac{\Delta H}{R}\frac{1}{T} - \frac{\Delta S}{R} \tag{41}$$

where R is the universal gas constant. ΔH denotes reaction enthalpy and ΔS the entropy change, which both can be determined experimentally. This enables the required equilibrium pressure to be determined for calculating the mass flow. Since the van't Hoff equation is a greatly simplified model for determining the equilibrium pressure, there are numerous methods for determining the equilibrium pressure more precisely [9].

Boundary conditions are required to complete the system, Eqs. (33)–(36). The metal hydride hydrogen storage tank is connected only from the left side to the network, see Fig. 1. Therefore, at the right end x_R of the space interval, which represents the bottom of the tank, conditions

$$u(x_R) = 0 , \quad \frac{\partial T}{\partial x}(x_R) = 0 \tag{42}$$

are applied. Due to Eq. (34), the first condition of (42) implies

$$\frac{\partial P}{\partial x}(x_R) = 0 . \tag{43}$$

At the inlet x_L of the tank, temperature $T(x_L)$ and pressure $P(x_L)$ are given by the connected node. Energy flux is $\dot{e}_L = c_v \dot{m}_L T(x_L)$ and mass flow is $\dot{m}_L = u(x_L)\rho_g(x_L)\epsilon A$. The gas density $\rho_g(x_L)$ can be computed from $T(x_L)$ and $P(x_L)$ using the ideal gas law, and velocity $u(x_L)$ is defined by (34).

3   Numerical Solution Approach

The numerical solution approach consists of four steps, explained below:

1. Semidiscretization of PDEs in space

When a constant diameter of the pipe is assumed, the hyperbolic PDE systems described by Eqs. (19)–(21) and (28)–(30) for gas and water flow through pipes can be summarized by

$$\frac{\partial q}{\partial t} + \frac{\partial}{\partial x} f(q) = S(x, t, q) , \quad x \in [x_L, x_R] \tag{44}$$

with state vector $q(x, t)$, flux function $f(q)$ and source term $S(x, t, q)$. For gas flow the state vector is given by $q = (\rho, u\rho, E)^T$ and for water flow by $q = (H, Q, T)^T$.

The space interval is discretized into $x_L = x_{1/2} < x_{3/2} < \ldots < x_{n+1/2} = x_R$ with constant stepsize $\Delta x = x_{i+1/2} - x_{i-1/2}$, cell centers $x_i = \frac{1}{2}(x_{i-1/2} + x_{i+1/2})$ and discrete state vectors $q_i(t) = q(x_i, t)$.

The space derivatives at the cells' center are approximated by $\frac{\partial}{\partial x} f(q_i) = \frac{f_{i+1/2} - f_{i-1/2}}{\Delta x}$ with suitable chosen flux values $f_{i\pm 1/2}$ leading to the semi-discretized system of ODEs

$$\frac{d}{dt} q_i(t) + \frac{f_{i+1/2} - f_{i-1/2}}{\Delta x} = S(x_i, t, q_i) , \quad i = 1, \ldots, n . \tag{45}$$

In order to get stable discretizations, a local Lax-Friedrichs approach is applied [17, 36]:

$$f_{i+1/2} = \frac{1}{2} \left(f_{i+1/2}^+ + f_{i+1/2}^- - |\lambda_{i+1/2}|(q_{i+1/2}^+ - q_{i+1/2}^-) \right) . \tag{46}$$

$\lambda_{i+1/2}$ denotes the eigenvalue of $\frac{df}{dq}$, which has locally around $x_{i+1/2}$ the largest absolute value. In the case of Euler equations this would be

$$|\lambda_{i+1/2}| = \max \left(|u_i| + \sqrt{\frac{\gamma P_i}{\rho_i}}, |u_{i+1}| + \sqrt{\frac{\gamma P_{i+1}}{\rho_{i+1}}} \right) .$$

A first order discretization is given by the choice

$$f_{i+1/2}^+ = f(q_{i+1}) , \quad f_{i+1/2}^- = f(q_i) , \quad q_{i+1/2}^+ = q_{i+1} , \quad q_{i+1/2}^- = q_i , \quad i = 0, \ldots, n$$

with linear extrapolated variables

$$q_0 = 2q_L - q_1 , \quad q_{n+1} = 2q_R - q_n . \tag{47}$$

The boundary values q_L, q_R are described in Sect. 2.2.4. Higher order can be achieved by WENO methods (see references [13, 26]).

In case of the hydride storage tank—described by Eqs. (33)–(36)—parabolic components are present. Moreover, the flow velocities are rather small such that a full discretization by central differences is sufficient.

For the implementation of the numerical boundary conditions of the Euler or water-hammer equations, the invariants given by Eqs. (24), (25) or (31), (32) are considered. First, a locally constant $\dot{x}$ is assumed and the invariants are continued to the boundary, here stated for the left side:

$$I_L(x_L, t_1) := I_L(x_1, t) \quad \text{with} \quad t_1 = t - \frac{\Delta x}{2\dot{x}} ,$$

$$I_L(x_L, t_2) := I_L(x_2, t) \quad \text{with} \quad t_2 = t - \frac{3\Delta x}{2\dot{x}} .$$

Next, $I_L(x_L, t_1)$, $I_L(x_L, t_2)$ are linear extrapolated to time t, resulting in the boundary conditions, given for the left and right side

$$I_L(x_L, t) = \frac{3}{2} I_L(x_1, t) - \frac{1}{2} I_L(x_2, t) , \tag{48}$$

$$I_R(x_R, t) = \frac{3}{2} I_R(x_n, t) - \frac{1}{2} I_R(x_{n-1}, t) . \tag{49}$$

As an alternative to the invariants I_L, I_R, the mass flow variables $\dot{m}(x_1, t)$, $\dot{m}(x_2, t)$ and $\dot{m}(x_n, t)$, $\dot{m}(x_{n-1}, t)$ may also be extrapolated to the boundaries x_L and x_R.

The implementation of the boundary conditions for the hydrogen storage tank is straight forward when using central finite differences and extrapolated variables according to Eq. (47).

2. Defining a DAE-system

In the next step, all equations for the different nodes and edges are combined into one system. Coupling conditions, like Eqs. (1)–(6), or for connections, like Eqs. (12) and (13), are linear equations. Pumps or valves are modelled by non-linear equations, for example Eqs. (15) and (18). Numerical boundary conditions, like Eqs. (48) and (49), are generally non-linear too since the invariants may depend non-linearly on the state variables (e.g. Eqs. (24) and (25)). Nodes representing tanks are defined by Eqs. (7)–(10). In this case it is preferable to change the state values of such nodes to mass M and mass times temperature MT, instead of pressure P and temperature T. Then Eqs. (7)–(10) are linear ODEs. Finally, the semi-discretized pipe Eq. (45) and the equations for the hydrogen storage tank are non-linear ODEs.

Collecting all state variables in a large vector $y(t)$ yields a monolithic DAE-system of type

$$My' = f(t, y) , \quad y(t_0) = y_0 , \tag{50}$$

where M is a constant diagonal matrix with entries one, when the corresponding state variable occurs on the left hand-side of an ordinary differential equation, and entries zero otherwise.

3. Computation of consistent initial values

In order to solve Eq. (50), consistent initial values $y(t_0) = y_0$ at initial time t_0 must be provided. y_0 is consistent, when all algebraic equations belonging to entries zero on the diagonal of M are fulfilled. One possibility would be the calculation of a steady state solution $0 = f(t_0, y_0)$ of the whole system (Eq. (50)). But such a solution might not exist. This can easily be explained as follows. When a water network is considered and the water input by source terms does not coincide with water consumption at time t_0, resulting in no solution. Furthermore, a steady state solution might have no physical relevance.

Therefore, the following approach is preferred: At initial time t_0, all pressure variables and temperature variables in the whole network are set to constant values

$P_i = P_0$, $T_i = T_0$ $\forall i$ and all mass and energy fluxes are set to zero $\dot{m}_j = 0$, $\dot{e}_j = 0$ $\forall j$. By doing so, most of the algebraic equations are fulfilled. Exceptions, like Eq. (12), are modified according to

$$P_R - P_L = t_{scale}\Delta P \tag{51}$$

by introducing a scaling time $t_{scale} = \min(1, \frac{t-t_0}{t_s})$ with a relaxation time t_s. Within this time period the algebraic equation is adapted from a consistent one to a physical relevant equation. Although all algebraic equations are now fulfilled at time $t = t_0 + t_s$, the initial state of the network might not match measurements. For example, in case of a water network the water levels in the storage tanks will not correspond to measured values and therefore incorrect pressure conditions will occur throughout the network. A possible solution to this problem is the usage of the external sources $\dot{m}_{s,i}$ in Eqs. (7)–(10), allowing one to adjust the simulated water levels to the measured ones. An example is given in Eq. (77) of Sect. 4.

4. Solution of the DAEs by ROW methods

Finally, the DAE-system Eq. (50) must be solved. In general, any suitable solver can be used. Special attention should be paid to the robustness of the methods. In the practical applications discussed here, aspects such as very high accuracy are often less important than obtaining reliable results. ROW methods have proven to be good candidates for such problems in many applications (see references [14, 20, 22]). Such methods for DAEs were first introduced by Roche [23] for autonomous systems of the form

$$y' = f(y, z), \quad y(t_0) = y_0, \tag{52}$$

$$0 = g(y, z), \quad z(t_0) = z_0. \tag{53}$$

It is assumed that $\frac{\partial g}{\partial z}$ is non singular (also called index-1 assumption). ROW methods applied to problems given by Eq. (50) are defined by (see reference [8]):

$$y_1 = y_0 + \sum_{i=1}^{s} b_i k_i, \quad i = 1, ..., s \tag{54}$$

$$(M - h\gamma J_0) k_i = hf(t_0 + \alpha_i h, y_0 + \sum_{j=1}^{i-1} \alpha_{ij} k_j) + hJ_0 \sum_{j=1}^{i-1} \gamma_{ij} k_j + h^2 \gamma_i \frac{\partial f}{\partial t}(t_0, y_0). \tag{55}$$

y_1 denotes an approximation to the solution of Eq. (50) at time $t = t_0 + h$. The method's coefficients are $\gamma, \alpha_{ij}, \gamma_{ij}, i = 1, \ldots, s, j = 1, \ldots, i - 1$ with $\alpha_i = \sum_{j=1}^{i-1} \alpha_{ij}, \gamma_i = \gamma + \sum_{j=1}^{i-1} \gamma_{ij}$ and the weights are b_i. The Jacobian matrix is evaluated at time $t = t_0$ (i.e. $J_0 = \frac{\partial f}{\partial y}(t_0, y_0)$).

An advantage of these methods is that at each time step only s linear equation systems with identical matrices have to be solved, where s is the stage number. A disadvantage compared to, for example, implicit multi-step or Runge-Kutta methods is that the Jacobian matrix must be recalculated at each time step.

There are a number of well known ROW methods for DAEs (see chapter by J. Lang and references therein). From these, the methods `ros3prl2` [21] and `rodasp` [27] were selected. Both schemes are embedded methods of order 3(2) and 4(3), respectively. The embedding enables an error estimator and thus automatic step size control. Furthermore, both methods are stiffly accurate and were constructed on the basis of `ros3pl` [15] and `rodas` [8]. They fulfill additional order conditions, required to avoid order reduction for the Prothero-Robinson equation and parabolic problems.

In addition to `ros3prl2` and `rodasp`, the standard integrators `ode15s`, `ode23t`, `ode15i` of MATLAB [25] and implicit Runge-Kutta method `radau5` [8] are applied in the numerical tests.

4 Numerical Examples

In the following, we provide three numerical examples—two for water supply network modelling and one for a hydrogen flow problem. The first example is very simple, but demonstrates some numerical difficulties arising in practical applications. It is shown that a proper numerical simulation requires the preliminary smoothing of input data and a robust integrator. The second example serves as a benchmark problem for DAE solvers. In both problems, energy fluxes are not considered. The third example shows the applicability of the introduced modelling approach in particular for hydrogen storage in metal hydride.

1. Pumping between two tanks

We start with a system consisting of two tanks and a pump in between, as seen in Fig. 4. The pump transports water from tank 1 to tank 2. Simultaneously, water is removed from tank 2 by a given volume flow $Q_{out}(t)$. A pump is described by its characteristic curve, which is the relationship between delivery head H and volume flow Q (see the left-hand-side subfigure within Fig. 5).

A typical example of such a curve is

$$H(Q) = \alpha_0 - \alpha_r Q^r \tag{56}$$

Fig. 4 Pumping between two tanks

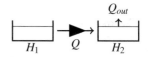

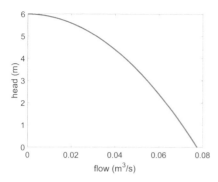
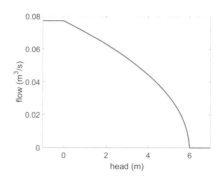

Fig. 5 Pump curves. Left $H(Q)$, right $Q(H)$ with extrapolation

with given positive coefficients α_0, α_r and coefficient $r > 1$. Since improper physical conditions for pump usage can occur during a simulation, extrapolation of the pump curve is required. Therefore, it is advantageous to extrapolate the inverse curve

$$Q(H) = \left(\frac{\alpha_0 - H}{\alpha_r}\right)^{1/r}, \tag{57}$$

shown in right-hand-side of Fig. 5. In the given example below, the pump operates around the maximum pumping head $H = \alpha_0$, where the function is not differentiable and a singularity in its first derivative occurs. Numerical tests have shown that Eq. (57) should be replaced by

$$Q(H) = \frac{(\alpha_0 - H)^2}{\alpha_r^{1/r}}(\alpha_0 - H + \epsilon)^{1/r-2} \tag{58}$$

with a small parameter ϵ, see [33]. Now, the extrapolated curve is continuously differentiable at $H = \alpha_0$. Regarding volume conservation of water, the model of the system shown in Fig. 4 is given by equations

$$H_1'(t) = -\frac{1}{A_1}Q(t), \tag{59}$$

$$H_2'(t) = \frac{1}{A_2}(Q(t) - Q_{out}(t)), \tag{60}$$

$$0 = Q(t) - i(t)\frac{(\alpha_0 - \Delta H)^2}{\alpha_r^{1/r}}(\alpha_0 - \Delta H + \epsilon)^{1/r-2}, \tag{61}$$

with water tank elevations of $H_1(t)$ and $H_2(t)$, and given tank base areas of A_1 and A_2. Extrapolation of the pump curve is implemented by the delivery head $\Delta H = \min(\max(H_2(t) - H_1(t), 0), \alpha_0)$, and function $i(t)$ indicates when the pump is on

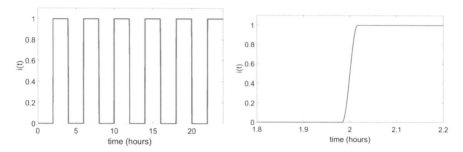

Fig. 6 Total pump schedule (left), zoomed view (right) that shows smoothing by Eq. (62)

($i(t) = 1$) or off ($i(t) = 0$). Since the network approach requires modelling by DAEs, system (59)–(61) are treated as a DAE system too.

For this example, the following data was used: $\epsilon = 10^{-2}$, $Q_{out}(t) = 0.0005\,(1 + \sin(\frac{2\pi t}{4 \cdot 3600}))$, $A_1 = 3$, $A_2 = 4$, $\alpha_0 = 6$, $\alpha_r = 1000$, $r = 2$ and a time interval of $t \in [0, 24 \cdot 3600]$ was simulated. When the pump is off at initial time $t = 0$, consistent initial values are given by $H_1(0) = H_2(0) = 10$, $Q(0) = 0$. The pump schedule shown in Fig. 6 was chosen. After 2 h (i.e. $t = 7200$ s), the pump is switched on, followed by alternating its on/off state every 2 h.

At switching times, $i(t)$ should be smoothed in order to avoid discontinuities. This smoothing near switch times t_j is done by

$$i(t) = \frac{1}{2}\sin(\pi \frac{t - t_j + t_s}{2t_s} - \frac{\pi}{2}) + \frac{1}{2} \quad \text{for} \quad t \in [t_j - t_s, t_j + t_s]. \tag{62}$$

In this case, switching in time interval $[t_j - t_s, t_j + t_s]$ from $i(t_j - t_s) = 0$ to $i(t_j + t_s) = 1$ is assumed, see Fig. 6. Switching from $i = 1$ to $i = 0$ is implemented analogously. The time period $t_s = 60$ for defining the length of the switching interval has been chosen in the numerical tests.

The resulting behaviour is shown in Fig. 7. Due to the extraction of water by $Q_{out}(t)$ the mean water levels in both tanks decrease. At each time when the pump is switched on, the water level of tank 1 drops and that of tank 2 rises rapidly and a sharp peak in the flow is generated. By zooming into the curve, one can see that the pump generates a small flow during its entire operating time.

In an initial test, the integrators `ode15s`, `ode23t`, `ode15i` of MATLAB, implicit Runge-Kutta method `radau5` and Rosenbrock methods `ros3prl2`, `rodasp` are applied to the problem. Results obtained with the usual tolerances $rtol = atol = 10^{-4}$ and 10^{-6} are given in Table 1. Although MATLAB's multistep procedures require more time steps, their computing times are the shortest since a lower number of Jacobian matrices evaluations are needed in comparison to the one-step methods.

Next the pump curve is changed slightly by setting the coefficient $r = 4$ and changing the smoothing parameter for the pump to $\epsilon = 10^{-3}$. Again, integrators are ran with a tolerance $tol = 10^{-4}$. Figure 8 shows that the multistep methods

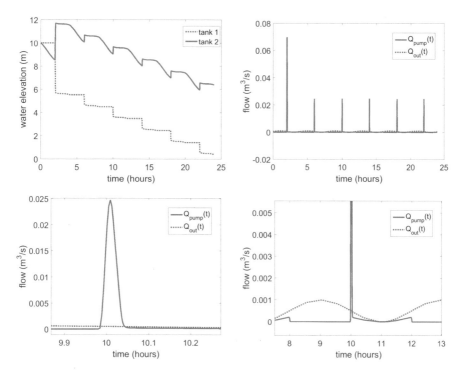

Fig. 7 Resulting model behavior, as given by Eqs. (59)–(61). Upper left: water elevations $H_1(t)$ and $H_2(t)$, upper right and lower: flows $Q(t)$ (denoted as $Q_{pump}(t)$) and $Q_{out}(t)$, together with zooms

Table 1 Numerical results for pumping problem. NSTEP = number of successful time steps, NFAIL = number of rejected time-steps, NJAC = number of Jacobian evaluations, NFCN = number of evaluations of the right-hand side of equation (50), CPU = computing time, ERR = maximum absolute error at t_{end}

Method	tol	NSTEP	NFAIL	NJAC	NFCN	CPU	ERR
ode23t	10^{-4}	464	165	63	1499	0.15	3.6e−3
ode15s		508	167	65	1581	0.15	4.6e−4
ode15i		593	144	48	1661	0.18	3.9e−3
radau5		227	16	185	4061	0.40	1.7e−6
ros3prl2		288	154	288	2324	0.30	7.1e−4
rodasp		202	97	202	2505	0.26	7.5e−5
ode23t	10^{-6}	1612	121	51	3650	0.34	7.6e−4
ode15s		1097	255	64	2820	0.28	4.1e−4
ode15i		1572	236	58	3413	0.41	3.7e−4
radau5		301	34	248	5060	0.68	4.5e−7
ros3prl2		1527	173	1527	11,035	1.26	6.7e−5
rodasp		450	127	450	3462	0.51	2.6e−5

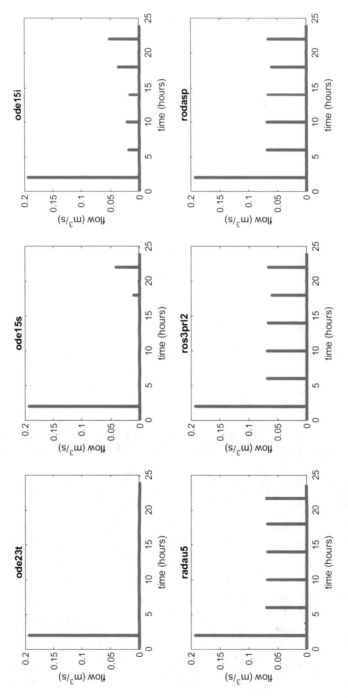

Fig. 8 Resulting model behavior with changed pump curve

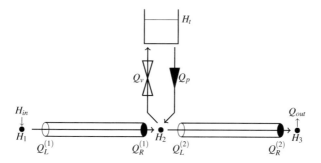

Fig. 9 A tank and valve system

lose accuracy. When reducing *tol* further they may even fail to solve the problem over the entire time interval. Only if the assigned tolerance is below the value of $tol = 10^{-6}$ they will deliver reliable results again.

This simple example demonstrated that a robust integration method is very important, especially if the simulation model is applied operationally within a forecast or optimization software. The operators in the control stations are not able to evaluate the cause of program terminations or adjust parameters such as tolerance.

2. Tank and valve system

The second example, shown in Fig. 9, is slightly more complicated. A predefined pressure $H_{in}(t)$ is introduced into the system via node 1. This node could represent the outlet of a waterworks. At a waterworks' outlet, a defined pressure is usually provided to deliver water to the supply area. Water is extracted from the system at node 3 using term $Q_{out}(t)$. Therefore, node 3 should represent the supply area. Water can be stored temporarily in the tank when the valve, represented in Fig. 9 as the edge in the direction of the tank, is open. If pressure in the supply area drops too far, the valve is closed and water from the reservoir is pumped into the supply area to increase pressure. Nodes 1, 2 and 2, 3 are connected by pipes. The whole DAE-system for modelling this water system is given by Eqs. (63)–(76).

$$0 = H_1 - H_{in} \tag{63}$$

$$0 = 2\left(gA_1H_1 - aQ_L^{(1)}\right) - 3\left(gA_1H_1^{(1)} - aQ_1^{(1)}\right) + \left(gA_1H_2^{(1)} - aQ_2^{(1)}\right) \tag{64}$$

$$(H_i^{(1)})' = -\frac{a^2}{gA_1}\frac{Q_{i+1/2}^{(1)} - Q_{i-1/2}^{(1)}}{\Delta x} + a\frac{H_{i+1/2}^{(1)} - H_{i-1/2}^{(1)}}{\Delta x}, \quad i = 1, ..., n \tag{65}$$

$$(Q_i^{(1)})' = -gA_1\frac{H_{i+1/2}^{(1)} - H_{i-1/2}^{(1)}}{\Delta x} + a\frac{Q_{i+1/2}^{(1)} - Q_{i-1/2}^{(1)}}{\Delta x} - \lambda(Q_i^{(1)})\frac{Q_i^{(1)}|Q_i^{(1)}|}{2D_1A_1}, \tag{66}$$

$$i = 1, ..., n$$

$$0 = 2\left(gA_1H_2 + aQ_R^{(1)}\right) - 3\left(gA_1H_n^{(1)} + aQ_n^{(1)}\right) + \left(gA_1H_{n-1}^{(1)} + aQ_{n-1}^{(1)}\right) \tag{67}$$

$$0 = Q_R^{(1)} + Q_p - Q_L^{(2)} - Q_v \tag{68}$$

$$0 = 2\left(g A_2 H_2 - a Q_L^{(2)}\right) - 3\left(g A_2 H_1^{(2)} - a Q_1^{(2)}\right) + \left(g A_2 H_2^{(2)} - a Q_2^{(2)}\right) \tag{69}$$

$$(H_i^{(2)})' = -\frac{a^2}{g A_2} \frac{Q_{i+1/2}^{(2)} - Q_{i-1/2}^{(2)}}{\Delta x} + a \frac{H_{i+1/2}^{(2)} - H_{i-1/2}^{(2)}}{\Delta x}, \quad i = 1, ..., n \tag{70}$$

$$(Q_i^{(2)})' = -g A_2 \frac{H_{i+1/2}^{(2)} - H_{i-1/2}^{(2)}}{\Delta x} + a \frac{Q_{i+1/2}^{(2)} - Q_{i-1/2}^{(2)}}{\Delta x} - \lambda(Q_i^{(2)}) \frac{Q_i^{(2)} |Q_i^{(2)}|}{2 D_2 A_2}, \tag{71}$$

$$i = 1, ..., n$$

$$0 = 2\left(g A_2 H_3 + a Q_R^{(2)}\right) - 3\left(g A_2 H_n^{(2)} + a Q_n^{(2)}\right) + \left(g A_2 H_{n-1}^{(2)} + a Q_{n-1}^{(2)}\right) \tag{72}$$

$$0 = Q_R^{(2)} - Q_{out} \tag{73}$$

$$0 = Q_p - i(t) \frac{(\alpha_0 - (H_2 - H_t))^2}{\alpha_r^{1/r}} (\alpha_0 - (H_2 - H_t) + \epsilon_p)^{1/r-2} \tag{74}$$

$$0 = Q_v \sqrt{\zeta(|H_2 - H_t| + \epsilon_v)} - s(t)(H_2 - H_t) \tag{75}$$

$$H_t' = \frac{1}{A_t}(Q_v - Q_p + Q_s) \tag{76}$$

The state vector $y(t)$ of this system is given by

$$y = (H_1, Q_L^{(1)}, H_1^{(1)}, Q_1^{(1)}, ..., H_n^{(1)}, Q_n^{(1)}, Q_R^{(1)}, H_2, Q_L^{(2)}, H_1^{(2)}, Q_1^{(2)}, ...,$$

$$H_n^{(2)}, Q_n^{(2)}, Q_R^{(2)}, H_3, Q_v, H_t, Q_p)^T$$

Here H_1, H_2, H_3, H_t are the pressure heads in the nodes and the tank, while Q_v and Q_p are volume flows through the valve and the pump. $Q_L^{(1)}, Q_R^{(1)}, Q_L^{(2)}$ and $Q_R^{(2)}$ are in- and outflows of pipes 1 and 2, while $H_i^{(j)}$ and $Q_i^{(j)}$ ($i = 1, ..., n$, $j \in \{1, 2\}$) are the inner pressure and flow states of the pipes, which are discretized into n space points.

Using Eq. (63), the given input pressure is defined, while Eqs. (68) and (73) describe the conservation of volume flow at nodes 2 and 3. Flow through the pump is defined by Eq. (74). The storage of water in the tank is described by Eq. (76). $Q_s(t)$ is an external in- or outflow of the tank defined by

$$Q_s(t) = A_t \frac{H_0 - H_t(t)}{t_{re}}. \tag{77}$$

This term causes the water level in the storage tank to approach the target value H_0, and t_{re} is the relaxation time. Normally, the external source term is only taken into account in the first phase of the simulation in order to adapt the simulated water levels in the storage tank to measured values. In this application, Q_s is applied during the first 8 h of simulation time with values $H_0 = 80$ and $t_{re} = 100$. Equation (75) provides the flow Q_v through the valve. Its square is proportional to the pressure loss. The valve coefficient is given by $\zeta = 1$, and $\epsilon_v = 0.05$ is a smoothing parameter. The opening degree of the valve is governed by $s(t) \in [0, 1]$. Here, only discrete values $s \in \{0, 1\}$ are assumed and smoothing is done in the same way as the switch function $i(t)$ of a pump.

Finally, Eqs. (65), (66), (70) and (71) are the semidiscretizations of the water hammer equations, and (64), (67), (69) and (72) describe the numerical boundary conditions. In order to make the model comprehensible, only first-order approxima- tions are used for the space derivatives (i.e. $H_{i+1/2} = \frac{1}{2}(H_i + H_{i+1})$, $Q_{i+1/2} = \frac{1}{2}(Q_i + Q_{i+1})$ with $H_0 = H_L$, $Q_0 = Q_L$ and $H_{n+1} = H_R$, $Q_{n+1} = Q_R$). The friction coefficient λ is approximated by the Swamee-Jain-approximation of the Colebrook and White formula [12, 24]:

$$\lambda(Q) = \frac{0.25}{\left(\log_{10}\left(\frac{k}{3.7D} + \frac{5.74}{Re^{0.9}}\right)\right)^2} \tag{78}$$

with roughness parameter k of the pipe and the Reynolds number given by $Re = \frac{4|Q|}{\pi D v}$ with a kinematic viscosity v.

The following coefficients, all given in SI units, were chosen for this example: Gravitational constant $g = 9.81$, sound velocity $a = 1414$ and kinematic viscosity $v = 1.31 \cdot 10^{-6}$. A roughness parameter $k = 0.0005$, length $L_1 = L_2 = 5000$ and diameter $D_1 = D_2 = 0.5$ were given for the pipes. Each pipe is discretized into 100 points, leading to a total number of $neq = 410$ equations. Coefficients of the pump are $\alpha_0 = 50$, $\alpha_r = 4000$ and $r = 2$, with a tank base area of $A_t = 1000$.

The time span for simulation was 3 days (i.e. $t \in [0, 3 \cdot 86400]$). After the first 8 h, external flow $Q_s(t)$ was shut off. The initial values were chosen as zero flow $Q_0 = 0$ and a constant pressured head $H_0 = 52.5$ for all variables. Unfortunately, these values are not consistent because Eqs. (63), (73)–(75) are not fulfilled at initial time $t_0 = 0$. In order to prevent the simulator's failing, a scaling time $t_{scale} = \min(1, \frac{t}{10})$ was introduced. By multiplying the second summand of Eqs. (73)–(75) by t_{scale} and replacing Eq. (63) by $0 = H_1 - (1 - t_{scale})H_0 - t_{scale}H_{in}$, an initial transient phase of 10 s was introduced. Consequently, the initial values became consistent and the desired conditions were achieved within this time period.

The extracted flow at node 3 is

$$Q_{out}(t) = 0.15\left(1 + \sin(2\pi\frac{t - 7200}{12 \cdot 3600})\right)$$

and input pressure head is $H_{in} = 80$. During low water demand, each day from 8 am to 1 pm and from 8 pm to 1 am the valve is opened and water is entering the tank. During high water demand, each day from 2 am to 7 am and from 2 pm to 7 pm, the pump is operating and water from the tank is delivered back to the system. For the rest of the time the valve is closed and the pump is switched off. This behaviour is well demonstrated by the simulation results shown in Fig. 10. Table 2 shows the numerical efforts for methods rodasp and radau5. Despite

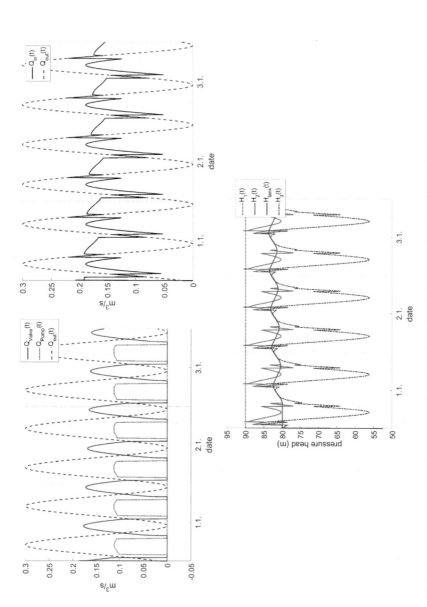

Fig. 10 Solution behaviour: Flow through valve, pump and outflow (left, above), inflow and outflow (right, above) and pressure heads at nodes and tank (below)

Table 2 Numerical results for second example problem. NSTEP = number of successful time steps, NFAIL = number of rejected time-steps, NJAC = number of Jacobian evaluations, NFCN = number of evaluations of the right-hand side, CPU = computing time

Method	tol	NSTEP	NFAIL	NJAC	NFCN	CPU
radau5	10^{-4}	2767	133	1419	36,943	7.9
ros3prl2	10^{-4}	4061	1015	4061	42,640	11.0
rodasp	10^{-4}	3311	788	3311	46,983	11.7
radau5	10^{-5}	Failed				
ros3prl2	10^{-5}	10,768	2401	10,768	112,482	27.8
rodasp	10^{-5}	6757	1672	6757	96,201	23.1

Fig. 11 Test configuration consisting of the pressure tank, that is connected to the hydride tank. The hydride tank has a thermal connection to a heat exchanger

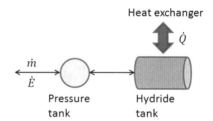

radau5 is more efficient, it is not able to solve the system with tolerances lower than $tol = 10^{-4}$. Moreover, integrator ode15s, ode23t, ode15i fail as well for all tested tolerances. Again, rodasp proves to be a very robust scheme.

Therefore, rodasp has been chosen as the standard integrator in many applications. It should be noted that the authors do not claim, that rodasp is the most efficient or accurate, but it has proved to be very robust. There are also many cases where rodasp is outperformed by radau5 with respect to efficiency.

3. Hydride storage

The third numerical example treats a hydride storage that is modeled as an edge with only one connected node. Thus, the gas can only flow in or out from one direction. The hydride storage is connected to a pressure tank (Eq. (7)) that supplies or removes the tank's gas such that the absorbing or desorbing process takes place. The example can be conceptualized by Fig. 11.

The model consists of Eq. (7) for the pressure tank, Eq. (37) for the heat exchanger and Eqs. (33)–(36) for the hydride tank. The pressure and hydride tanks have volumes of $V = 0.005$ and $V = 0.000326$ m^3, respectively. The pressure tank is constantly refilled to hold a nearly constant mass of $m_{target} = 0.0061$ kg for absorption, which is given at a pressure of $p = 15$ bar and a temperature of $T = 298$ K according to the ideal gas law. In the case of desorption the mass is decreased to $m_{target} = 0.00041$ kg at $T = 298$ K and $p = 1$ bar. The following equation is used for the refilling process:

$$\dot{m}_{in} = \frac{m_{target} - m_{tank}}{t_{re}} \tag{79}$$

where m_{tank} is the actual mass in the tank and t_{re} is the refilling time. The energy flux can be calculated by

$$\dot{E}_{in} = c_v T_{in} \frac{m_{target} - m_{tank}}{t_{re}} \qquad (80)$$

with a constant inflow gas temperature of $T_{in} = 298$ K. To maintain the absorbing/desorbing process, the temperature of the hydride tank's solid phase is regulated by an fictive heat exchanger, which has a constant temperature of $T_a = 298$ K when absorbing or $T_a = 363$ K when desorbing. To keep it simple, this exchanger has a constant heat transfer coefficient, allowing Eq. (37) to be used for calculating the energy flow. All relevant simulation-parameters are given in Table 3.

For this model, simulation starts with absorption over $t_{abs} = 1100$ s and followed by desorption over $t_{des} = 1000$ s, and all results were obtained with the integrator rodasp. The resulting absorption and desorption mass loading and temperature as a function of simulation time are shown in Fig. 12. At the beginning of the absorbing process ($t \approx 0.05$ s) the reaction starts. There is a sharp increase of the mass load from 0 to 0.15% over 6 s. Due to the generated reaction energy, the temperature increases from 298 to 362.3 K. Because of the heat exchanger, the reaction continues and mass load increases from 0.15 to 1.39% with a smaller but constant slope until the saturation density is reached at $t \approx 1000$ s. The development of the temperature decreases from 362.3 K to 350 K over 780 s because of the decreasing reaction rate and the cooling by the heat exchanger. When the saturation density of the hydride is reached, the reaction stops and the slope of the temperature curve increase until

Table 3 Simulation parameters for modelling hydrogen storage in a metal hydride. For the thermal-physical properties of materials see references [11, 19]

Parameters	Value	Unit	Parameters	Value	Unit
ρ_{ss}, Saturated density s	8314	$\frac{kg}{m^3}$	ρ_{su}, Unsaturated density s	8200	$\frac{kg}{m^3}$
c_{pg}, Specific heat hydrogen	14,304	$\frac{J}{kg\,K}$	c_{ps}, Specific heat solid	419	$\frac{J}{kg\,K}$
λ_s, Thermal conductivity s	1.2	$\frac{W}{m^2\,K}$	R, Universal Gas constant	8.314	$\frac{J}{kg\,K}$
λ_g, Thermal conductivity g	0.1807	$\frac{W}{m^2\,K}$	ϵ, Porosity	0.5	–
E_a, Activation energy abs.	21,179.6	$\frac{J}{mol}$	γ, Ratio of specific heats	1.4	–
E_d, Activation energy des.	16,450	$\frac{J}{mol}$	M_g, Molecular weight gas	2.016	$\frac{g}{mol}$
C_a, Reaction constant abs.	59.187	$\frac{1}{s}$	ΔH, Enthalpy of reaction	30,780	$\frac{J}{mol}$
C_d, Reaction constant des.	9.57	$\frac{1}{s}$	ΔS, Entropy of reaction	107.2	$\frac{J}{mol\,K}$
ΔH, Enthalpy of reaction	30,780	$\frac{J}{mol}$	K, Permeability	10^{-8}	m^2
r, Radius of tank	0.036	m	h, High of tank	0.08	m
μ, Dynamic viscosity g	$8.813 \cdot 10^{-6}$	$\frac{kg}{ms}$	α, Heat transfer coefficient	160	$\frac{W}{m^2\,K}$

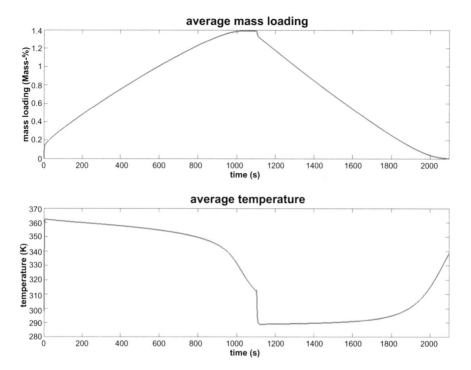

Fig. 12 Average mass loading and temperature profiles for the absorption and desorption process. The average mass loading of hydrogen is defined as $\frac{m_{H_2}}{m_{su}}$, where m_{su} is the mass of the unsaturated solid phase and m_{H_2} is the mass of gas

a temperature of 312 K at 1100 s is reached. After that the desorption starts and a similar but mirror-inverted curve characteristic can be seen.

The results show the desired behavior and are in good agreement to literature [11, 19]. Therefore, the hydride storage can be handled as a network element when setting up larger hydrogen networks for further applications.

5 Conclusion

The presented unified modelling approach for simulating flow networks seem well suited for many applications. In addition to the ones presented, it is an easy task to integrate further network elements as new edges. The numerical solution relies on a proper semidiscretization of the PDEs. Since conservation laws must be discretized, a local Lax-Friedrichs approach is used. Together with the proposed treatment of the boundary conditions, the method of lines leads to large DAE systems. Algebraic equations arise from the boundary conditions, the coupling conditions at nodes, and simple network elements (e.g pumps, valves). ODEs arise from the semidiscretized PDEs and storage nodes (e.g tanks).

Special emphasis must be paid to the computation of consistent initial values. For this task a simple approach was introduced. In practical applications it might be important to perform a preliminary smoothing to given data or characteristic curves. Examples of such problems were discussed in the first two numerical simulations. It was shown that Rosenbrock method rodasp is very robust and reliable, but was not the most efficient integrator in all cases.

In the final example, the integration of a metal hydride storage element into the presented network approach was presented. The integration of such a storage element demonstrates the flexibility of the overall methodology, and will allow for further complex applications to be done within the field of energy network modelling.

Acknowledgment The authors would like to thank the two reviewers and Karl Kirschner for many helpful suggestions for improvement.

References

1. J. Abreu, E. Cabrera, J. Izquierdo, J. García-Serra, Flow modelling in pressurized systems revisited. ASCE J. Hydraul. Eng. **125**(11), 1154–1169 (1999)
2. P. Bales, O. Kolb, J. Lang, *Hierarchical Modelling and Model Adaptivity for Gas Flow on Networks*. Lecture Notes in Computer Science 5544, Part I (Springer, Berlin, 2009), pp. 337–346
3. C. Bourdarias, S. Gerbi, A finite volume scheme for a model coupling free surface and pressurised flows in pipes. J. Comput. Appl. Math. **209**(1), 109–131 (2007)
4. H. Buchner, *Energiespeicherung in Metallhydriden* (Springer, Wien, 1982)
5. Y. Corapcioglu, *Advances in Porous Media*, vol. 3 (Elsevier, Amsterdam, 1996), p. 107
6. P. Domschke, B. Hiller, J. Lang, C. Tischendorf, Modellierung von Gasnetzwerken: Eine Übersicht. Technical Report SFBTRR 154, TU Darmstadt, 2017
7. E. Guelpa, A. Sciacovelli, V. Verda, Thermo-fluid dynamic model of large district heating networks for the analysis of primary energy savings. Energy (2017). ISSN 0360-5442, https://doi.org/10.1016/j.energy.2017.07.177
8. E. Hairer, G. Wanner, *Solving Ordinary Differential Equations II, Stiff and Differential Algebraic Problems*, 2nd ed. (Springer, Berlin, 1996)
9. K. Herbig, L. Rötzsch, C. Pohlmann, T. Weißgärber, B. Kieback, Hydrogen storage systems based on hydride–graphite composites: computer simulation and experimental validation. Int. J. Hydrog. Energy **38**, 7026–7036 (2013)
10. C. Hirsch, *Numerical Computation of Internal and External Flows*, vol. 2 (Wiley, Chichester, 1984)
11. V. Keshari, M. Maiya, Design and investigation of hydriding alloy based hydrogen storage reactor integrated with a pin fin tube heat exchanger. Int. J. Hydrog. Energy **43**(14), 7081–7095 (2018)
12. O. Kolb, *Simulation and Optimization of Gas and Water Supply Networks* (Verlag Dr. Hut, München, 2011)
13. A. Kurganov, D. Levy, A third-order semidiscrete central scheme for conservation laws and convection-diffusion equations. SIAM J. Sci. Comput. **22**(4), 1461–1488 (2000)

14. J. Lang, B. Erdmann, *Adaptive Linearly Implicit Methods for Heat and Mass Transfer*, ed. by A.V. Wouver, P. Saucez, W.E. Schiesser. Adaptive Method of Lines (CRC Press, Boca Raton, 2000), pp. 295–316
15. J. Lang, D. Teleaga, Towards a fully space-time adaptive FEM for magnetoquasistatics. IEEE Trans. Magn. **44**, 1238–1241 (2008)
16. B. Lendt, G. Cerbe, *Grundlagen der Gastechnik* (Hanser Verlag, München, 2016)
17. R.J. LeVeque, *Finite Volume Methods for Hyperbolic Problems* (Cambridge University Press, Cambridge, 2002)
18. S. Mohammadshahi, E. Gray, C. Webb, A review of mathematical modelling of metal-hydride systems for hydrogen storage applications. Int. J. Hydrog. Energy **41**, 3470–3484 (2016)
19. P. Muthukumar, S. Venkata Ramana, Numerical simulation of coupled heat and mass transfer in metal hydride-based hydrogen storage reactor. J. Alloys Compd. **472**, 466–472 (2008)
20. J. Rang, A new stiffly accurate Rosenbrock-Wanner method for solving the incompressible Navier-Stokes equations, in *Notes on Numerical Fluid Mechanics and Multidisciplinary Design*, vol. 120 (2013). https://doi.org/10.1007/978-3-642-33221-0_18
21. J. Rang, Improved traditional Rosenbrock–Wanner methods for stiff ODEs and DAEs. J. Comput. Appl. Math. **286**, 128–144 (2015). https://doi.org/10.1016/j.cam.2015.03.010
22. P. Rentrop, G. Steinebach, Model and numerical techniques for the alarm system of river Rhine. Surv. Math. Ind. **6**, 245–265 (1997)
23. M. Roche, Rosenbrock methods for differential algebraic equations. Numer. Math. **52**, 45–63 (1988)
24. L.A. Rossman, *EPANET 2 Users Manual* (U.S. Environmental Protection Agency, Cincinnati, 2000)
25. L.F. Shampine, I. Gladwell, S. Thompson, *Solving ODEs with MATLAB* (Cambridge University Press, Cambridge, 2003)
26. C.-W. Shu, High order weighted essentially nonoscillatory schemes for convection dominated problems. SIAM Rev. **51**(1), 82–126 (2009)
27. G. Steinebach, Order-reduction of ROW-methods for DAEs and method of lines applications. Preprint-Nr. 1741, FB Mathematik, TH Darmstadt, 1995
28. G. Steinebach, From river Rhine alarm model to water supply network simulation by the method of lines, in *Progress in Industrial Mathematics at ECMI 2014*, ed. by G. Russo, V. Capasso, G. Nicosia, V. Romano (Springer, Berlin, 2017), pp. 783–792
29. G. Steinebach, P. Rentrop, An adaptive method of lines approach for modelling flow and transport in rivers, in *Adaptive Method of Lines*, ed. by A.Vande Wouver, Ph. Sauces, W.E. Schiesser (Chapman & Hall/CRC, Boca Raton, 2001), pp. 181–205
30. G. Steinebach, K. Wilke, Flood forecasting and warning on the River Rhine. Water and environmental management. J. CIWEM **14**(1), 39–44 (2000)
31. G. Steinebach, M. Paffrath, P. Rentrop, R. Rosen, S. Seidl, Process simulation for sewer systems by splitting. Int. J. Eng. Syst. Model. Simul. **1**(4), 185–192 (2009)
32. G. Steinebach, R. Rosen, A. Sohr, Modelling and numerical simulation of pipe flow problems in water supply systems, in *Mathematical Optimization of Water Networks*, ed. by A. Martin, K. Klamroth, J. Lang, G. Leugering, A. Morsi, M. Oberlack, M. Ostrowski, R. Rosen. International Series of Numerical Mathematics, vol. 162 (2012), pp. 3–15
33. G. Steinebach, T. Jax, P. Hausmann, D. Dreistadt, *ERWAS - Verbundprojekt EWave: Energiemanagementsystem Wasserversorgung - Abschlussbericht zu Teilprojekt 5 -*. Bundesministerium für Bildung und Forschung (2018)
34. J.J. Stoker, *Water Waves, the Mathematical Theory with Applications* (Interscience, New York, 1957)
35. A. Sutiono, Modellierung und Simulation einer thermochemischen Kältemaschine auf Metallhydridbasis. Dissertation, Technischen Universität Hamburg-Harburg, 2011
36. E.F. Toro, *Riemann Solvers and Numerical Methods for Fluid Dynamics* (Springer, Berlin, 2009)

Exponential Rosenbrock Methods and Their Application in Visual Computing

Vu Thai Luan and Dominik L. Michels

1 Introduction

Developing numerical models for practical simulations in science and engineering usually results in problems regarding the presence of wide-range time scales. These problems involve both slow and fast components leading to rapid variations in the solution. This gives rise to the so-called *stiffness phenomena*. Typical examples are models in molecular dynamics (see e.g. [36]), chemical kinetics, combustion, mechanical vibrations (mass-spring-damper models), visual computing (specially in computer animation), computational fluid dynamics, meteorology, etc., just to name a few. They are usually formulated as systems of stiff differential equations which can be cast in the general form

$$u'(t) = F(u(t)), \quad u(t_0) = u_0, \tag{1}$$

where $u \in \mathbb{R}^n$ is the state vector and $F : \mathbb{R}^n \longrightarrow \mathbb{R}^n$ represents the vector field. The challenges in solving this system are due to its stiffness by means of the eigenvalues of the Jacobian matrix of F differing by several orders of magnitude. In the early days of developing numerical methods for ordinary differential equations (ODEs), classical methods such as the explicit Runge–Kutta integrators were proposed. For stiff problems, however, they are usually limited by stability issues due to the CFL condition leading to the use of unreasonable time steps, particularly for large-scale

V. T. Luan (✉)
Department of Mathematics and Statistics, Mississippi State University, Starkville, MS, USA
e-mail: luan@math.msstate.edu

D. L. Michels
Computational Sciences Group, Visual Computing Center, King Abdullah University of Science and Technology, Thuwal, Saudi Arabia
e-mail: dominik.michels@kaust.edu.sa

© The Author(s), under exclusive license to Springer Nature Switzerland AG 2021
T. Jax et al. (eds.), *Rosenbrock—Wanner–Type Methods*, Mathematics Online
First Collections, https://doi.org/10.1007/978-3-030-76810-2_3

applications. The introduction of implicit methods such as semi-implicit, IMEX (see [2]), and BDF methods (see [10, 14]) has changed the situation. Theses standard methods require the solution of nonlinear systems of equations in each step. As the stiffness of the problem increases, considerably computational effort is observed. This can be seen as a shortcoming of the implicit schemes.

In the last 20 years, with the new developments of numerical linear algebra algorithms in computing matrix functions [1, 25, 41], exponential integrators have become an alternative approach for stiff problems (see the survey [24]; next to physics simulations, exponential integrators are nowadays also employed for different applications as for the construction of hybrid Monte Carlo algorithms, see [7]). For the fully nonlinear stiff system (1), we mention good candidates, the so-called *explicit exponential Rosenbrock methods*, which can handle the stiffness of the system in an explicit and very accurate way. This class of exponential integrators was originally proposed in [23] and further developed in [26, 30, 32, 34]. They have shown to be very efficient both in terms of accuracy and computational savings. In particular, the lower-order schemes were recently successfully applied to a number of different applications [8, 15, 17, 46, 49] and very recently the fourth- and fifth-order schemes were shown to be the method of choice for some meteorological models (see [35]).

In this work, we show how the exponential Rosenbrock methods (particularly higher-order schemes) can be also applied efficiently in order to solve problems in computational modeling of elastodynamic systems of coupled oscillators (particle systems) which are often used in visual computing (e.g. for computer animation). In their simplest formulation, their dynamics can be described using Newton's second law of motion leading to a system of second-order ODEs of the form

$$m_i \ddot{x}_i + \sum_{j \in \mathcal{N}(i)} k_{ij}(\|x_i - x_j\| - \ell_{ij}) \frac{x_i - x_j}{\|x_i - x_j\|} = g_i(x_i, \dot{x}_i, \cdot), \quad i = 1, 2, \cdots, N,$$

(2)

where N is the number of particles, $x_i \in \mathbb{R}^3$, m_i, k_{ij}, ℓ_{ij} denote the position of particle i from the initial position, its mass, the spring stiffness, the equilibrium length of the spring between particles i and j, respectively, and $\mathcal{N}(i)$ denotes the set of indices of particles that are connected to particle i with a spring (the neighborhood of particle i). Finally, g_i represents the external force acting on particle i which can result from an external potential, collisions, etc., and can be dependent of all particle positions, velocities, or external forces set by user interaction.

Our approach for integrating (2) is first to reformulate it in the form of (1) (following a novel approach in [40]). The reformulated system is a very stiff one since the linear spring forces usually possess very high frequencies. Due to the special structure of its linear part (skew-symmetric matrix) and large nonlinearities, we then make use of exponential Rosenbrock methods. Moreover, we propose to use the improved algorithm in [35] for the evaluation of a linear combination of φ-functions acting on certain vectors $v_0, \ldots, v_p$, i.e. $\sum_{k=0}^{p} \varphi_k(A) v_k$ which is crucial for implementing exponential schemes. Our numerical results on a number of complex models in visual computing indicate that this approach significantly reduces computational time over the current state-of-the-art techniques while maintaining sufficient levels of accuracy.

This chapter is organized as follows. In Sect. 2, we present a reformulation of systems of coupled oscillators (2) in the form of (1) and briefly review previous approaches used for simulating these systems in visual computing. In Sect. 3, we describe the exponential Rosenbrock methods as an alternative approach for solving large stiff systems (1). The implementation of these methods is discussed in Sect. 4, where we also introduce a new procedure to further improve one of the state-of-the-art algorithms. In Sect. 5 we demonstrate the efficiency of the exponential Rosenbrock methods on a number of complex models in visual computing. In particular, we address the simulation of deformable bodies, fibers including elastic collisions, and crash scenarios including nonelastic deformations. These examples focus on relevant aspects in the realm of visual computing, like stability and energy conservation, large stiffness values, and high fidelity and visual accuracy. We include an evaluation against classical and state-of-the-art methods used in this field. Finally, some concluding remarks are given in Sect. 6.

2 Reformulation of Systems of Coupled Oscillators

We first consider the system of coupled oscillators (2). Let $x(t) \in \mathbb{R}^{3N}$, $M \in \mathbb{R}^{3N \times 3N}$, $D \in \mathbb{R}^{3N \times 3N}$, $K \in \mathbb{R}^{3N \times 3N}$ and $g(x) \in \mathbb{R}^{3N}$ denote the vector of positions, the mass matrix (often diagonal and thus nonsingular), the damping matrix, the spring matrix (stiff), and the total external forces acting on the system, respectively. Using these matrix notations and denoting $A = M^{-1}K$, (2) can be written as a system of second-order ODEs

$$x''(t) + Ax(t) = g(x(t)), \quad x(t_0) = x_0, \ x'(t_0) = v_0. \tag{3}$$

Here x_0, v_0 are some given initial positions and velocities. For simplicity we neglect damping and assume that A is a symmetric, positive definite matrix (this is a reasonable assumption in many models, see [38]). Therefore, there exists a unique positive definite matrix Ω such that $A = \Omega^2$ (and clearly Ω^{-1} exists).

Following our approach in [40], we introduce the new variable

$$u(t) = \begin{bmatrix} \Omega x(t) \\ x'(t) \end{bmatrix}. \tag{4}$$

Using this one can reformulate (3) as a first-order system of ODEs of the form like (1):

$$u'(t) = F(u(t)) = \mathscr{A}u(t) + G(u(t)), \quad u(t_0) = u_0, \tag{5}$$

where

$$\mathscr{A} = \begin{bmatrix} \mathbf{0} & \Omega \\ -\Omega & \mathbf{0} \end{bmatrix}, \quad G(u) = \begin{bmatrix} \mathbf{0} \\ g(x) \end{bmatrix}. \tag{6}$$

Since the linear spring forces usually possess high frequencies (thus $\|K\| \gg 1$ and so is $\|A\|$), (5) becomes a very stiff ODE. Regarding the new formulation (5)–(6), we observe the following two remarks.

Remark 1 Clearly, the new linear part associated with $\mathscr{A}$, that is a skew-symmetric matrix. We note that this significantly differs from the common way of reformulating (3) that is to use the change of variable $X(t) = [x(t), \ x'(t)]^T$ which results in a non-symmetric matrix. The advantage here is that since $\mathscr{A}$ is a skew-symmetric matrix, its nonzero eigenvalues are all pure imaginary and are in pairs $\pm\lambda_k i$. We also observe that $\mathscr{A}$ is an infinitesimal symplectic (or Hamiltonian). This is because, by definition of an infinitesimal symplectic matrix, we check whether $W\mathscr{A} + \mathscr{A}^T W = \mathbf{O}$, where W is the anti-symmetric matrix $W = \begin{bmatrix} \mathbf{0} & I \\ -I & \mathbf{0} \end{bmatrix}$. This can be easily verified since

$$W\mathscr{A} = \begin{bmatrix} -\Omega & \mathbf{0} \\ \mathbf{0} & -\Omega \end{bmatrix},$$

which is clearly a symmetric matrix, i.e., $W\mathscr{A} = (W\mathscr{A})^T$.

Remark 2 If the Jacobian matrix $F'(u) = \mathscr{A} + G'(u)$ is infinitesimal symplectic, (5) is a Hamiltonian system. This can be fulfilled since a typical situation in Hamiltonian systems is that $g(x) = \nabla f(x)$ for some function $f(x)$ and thus $g'(x) = \nabla^2 f(x)$ becomes a Hessian matrix, which is symmetric.

As seen, either using the common way (mentioned in Remark 1) or the new way (4) for reformulating (3), one has to solve the stiff ODE (5). In visual computing it is usually solved by explicit methods such as the fourth-order Runge–Kutta methods, semi-implicit methods such as the Störmer–Verlet methods, the backward differentiation formulas (BDF-1 and BDF-2) methods, or IMEX methods. In this regard, we refer to some contributions in the context of interacting deformable bodies, cloth, solids, and elastic rods, see [3, 4, 12, 16, 19, 47]. For large-scale applications associated with stiff systems, however, both types of these time integration techniques have their own limitations as mentioned in the introduction. In recent years, exponential integrators have shown to be competitive for large-scale problems in physics and for nonlinear parabolic PDEs, as well as for highly oscillatory problems (see [24]). They have attracted much attention by the broad computational mathematics community since mid-1990s. At the time while solving linear systems $(I - \alpha hJ)x = v$ with some Jacobian matrix J (required when using implicit methods) is generally only of linear convergence, it was realized that Krylov subspace methods for approximating the action of a matrix exponential on a vector, $e^{hJ}v$, offer superlinear convergence (see [21]). Unless a good preconditioner is available, this is clearly a computational advantage of exponential integrators over implicit methods. This has been addressed in the visual computing community very recently through a number of interesting work on exponential integrators, see e.g.[37–40]. Inspired by this interest, in the following sections we will show how exponential Rosenbrock methods—one of the popular classes of exponential integrators—can be applied for simulating systems of coupled oscillators.

3 Explicit Exponential Rosenbrock Methods

In this section, based on [23, 26, 29, 32, 34] we present a compact summary of the introduction of exponential Rosenbrock methods and their derivations for methods of order up to 5. We then display some efficient schemes for our numerical experiments for some applications in visual computing.

3.1 Approach

Motivated by the idea of deriving Rosenbrock-type methods, see [18, Chap. IV.7], instead of integrating the fully nonlinear system (1) (which has a large nonlinearity for stiff problems), one can replace it by a sequence of semilinear problems

$$u'(t) = F(u(t)) = J_n u(t) + g_n(u(t)), \tag{7}$$

by linearizing the forcing term $F(u)$ in each time step at the numerical solution u_n (due to [42]) with

$$J_n = F'(u_n), \quad g_n(u) = F(u) - J_n u \tag{8}$$

are the Jacobian and the nonlinear remainder, respectively. An advantage of this approach is that $g'_n(u_n) = F'(u_n) - J_n = 0$ which shows that the new nonlinearity $g_n(u)$ has a much smaller Lipschitz constant than that of the original one $F(u)$. The next idea is to handle the stiffness by solving the linear part $J_n u$ exactly and integrating the new nonlinearity $g_n(u)$ explicitly. For that, the representation of the exact solution at time $t_{n+1} = t_n + h$ of (7) using the variation-of-constants formula

$$u(t_{n+1}) = e^{hJ_n}u(t_n) + \int_0^h e^{(h-\tau)J_n} g_n(u(t_n + \tau))d\tau \tag{9}$$

plays a crucial role in constructing this type of integrators. As seen from (9), while the linear part can be integrated exactly by computing the action of the matrix exponential e^{hJ_n} on the vector $u(t_n)$, the integral involving $g_n(u)$ can be approximated by some quadrature. This procedure results in the so-called *exponential Rosenbrock methods*, see [23, 26].

Remark 3 For the system of coupled oscillators (2), the forcing term $F(u)$ has the semilinear form (5), which can be considered as a fixed linearization. Therefore, one can directly apply explicit the exponential Runge–Kutta methods (see [22]) to (5). The advantage of these methods is that the time-step h is not restricted by the CFL condition when integrating the linear part $\mathscr{A}u$. In our applications, however, the nonlinearity $G(u)$ is large in which the CFL condition usually serves as a reference for setting the time-step. In particular, for the stability hL_G should be sufficiently small (L_G is the Lipschitz constant of $G(u)$). In this regard, the dynamic linearization approach (7) applied to (5)

$$u'(t) = F(u) = \mathscr{A}u + G(u) = J_n u + G_n(u) \tag{10}$$

with

$$J_n = \mathscr{A} + G'(u_n), \tag{11}$$

offers a great advantage in improving the stability (in each step) when integrating $G(u)$. This is because instead of integrating the original semilinear problem with large nonlinearity $G(u)$, we only have to deal with a much smaller nonlinearity $G_n(u)$ (as mentioned above). Note that the new linear part $J_n u$ with the Jacobian J_n now incorporates both $\mathscr{A}$ and the Jacobian of the nonlinearity $G(u)$, which can be again solved exactly. It is thus anticipated that this idea of exponential Rosenbrock methods opens up the possibility to take even larger time steps compared to exponential Runge–Kutta methods.

3.2   Formulation of a Second-Order and General Schemes

In this subsection, we will illustrate the approach of exponential Rosenbrock methods by presenting a simple derivation of a second-order scheme and formulating general schemes.

3.2.1   A Second-Order Scheme

First, expanding $u(t_n+\tau)$ in a Taylor series gives $u(t_n+\tau) = u(t_n)+\tau u'(t_n)+O(\tau^2)$. Then inserting this into $g_n(u(t_n + \tau))$ and again expanding it as a Taylor series around $u(t_n)$ (using $g'_n(u(t_n)) = 0$) leads to

$$g_n(u(t_n + \tau)) = g_n(u(t_n)) + O(\tau^2). \tag{12}$$

Inserting (12) into the integral of (9) and denoting $\varphi_1(h J_n) = \frac{1}{h} \int_0^h e^{(h-\tau)J_n} d\tau$ gives

$$u(t_{n+1}) = e^{h J_n} u(t_n) + h\varphi_1(h J_n) g_n(u(t_n)) + O(h^3). \tag{13}$$

Neglecting the local error term $O(h^3)$ results in a second-order scheme, which can be reformulated as

$$u_{n+1} = u_n + h\varphi_1(h J_n) F(u_n) \tag{14}$$

by replacing $g_n(u(t_n))$ by (8) and using the fact that $\varphi_1(z) = (e^z - 1)/z$. This scheme was derived before and named as *exponential Rosenbrock-Euler method*, see [23, 26] (since when considering the formal limit $J_n \to 0$, (14) is the underlying Euler method). The derivation here, however, shows directly that this scheme has an order of consistency three and thus it is a second-order stiffly accurate method (since the constant behind the Landau notation O only depends on the regularity assumptions on $u(t)$ and $g_n(u)$, but is independent of $\|J_n\|$).

3.2.2 General Schemes

For the derivation of higher-order schemes, one can proceed in a similar way as for the construction of classical Runge–Kutta methods. Namely, one can approximate the integral in (9) by using some higher-order quadrature rule with nodes c_i in $[0, 1]$ and weights $b_i(h J_n)$ which are matrix functions of $h J_n$, yielding

$$u(t_{n+1}) \approx e^{h J_n} u(t_n) + h \sum_{i=1}^{s} b_i(h J_n) g_n(u(t_n + c_i h)). \tag{15}$$

The unknown intermediate values $u(t_n + c_i h)$ can be again approximated by using (9) (with $c_i h$ in place of h) with another quadrature rule using the same nodes c_j, $1 \le j \le i - 1$, (to avoid generating new unknowns) and new weights $a_{ij}(h J_n)$, leading to

$$u(t_n + c_i h) \approx e^{c_i h J_n} u(t_n) + h_n \sum_{j=1}^{i-1} a_{ij}(h J_n) g_n(u(t_n + c_j h)). \tag{16}$$

Let us denote $u_n \approx u(t_n)$ and $U_{ni} \approx u(t_n + c_i h_n)$. As done for (14), using (12) (with $c_i h, h$ in place of τ, respectively) one can reformulate (15) and (16) in a similar manner, which yields the general format of s-stage explicit exponential Rosenbrock methods

$$
U_{ni} = u_n + c_i h \varphi_1(c_i h J_n) F(u_n) + h \sum_{j=2}^{i-1} a_{ij}(h J_n) D_{nj}, \tag{17a}
$$

$$
u_{n+1} = u_n + h \varphi_1(h J_n) F(u_n) + h \sum_{i=2}^{s} b_i(h J_n) D_{ni} \tag{17b}
$$

with

$$
D_{ni} = g_n(U_{ni}) - g_n(u_n), \tag{17c}
$$

As in (12), we have $D_{ni} = O(h^2)$. Thus, the general methods (17) are small perturbations of the exponential Rosenbrock–Euler method (14). Note that the weights $a_{ij}(h J_n)$ and $b_i(h J_n)$ are usually linear combinations of $\varphi_k(c_i h J_n)$ and $\varphi_k(h J_n)$, respectively, where the φ functions (similar to φ_1) are given by

$$
\varphi_k(h Z) = \frac{1}{h^k} \int_0^h e^{(h-\tau)Z} \frac{\tau^{k-1}}{(k-1)!} d\tau, \quad k \geq 1 \tag{18}
$$

and satisfy the recursion relation

$$
\varphi_{k+1}(z) = \frac{\varphi_k(z) - \frac{1}{k!}}{z}, \quad k \geq 1. \tag{19}
$$

It is important to note that these functions are bounded (uniformly) independently of $\|J_n\|$ (i.e. the stiffness) so are the coefficients $a_{ij}(h J_n)$ and $b_i(h J_n)$ (see e.g. [24]).

Clearly, using exponential Rosenbrock schemes (17) offers some good advantages. First, they do not require the solution of linear or nonlinear systems of equations. Second, as mentioned above, they offer a better stability when solving stiff problems with large nonlinearities and thus allow to use larger time-steps. Third, since the Jacobian of the new nonlinearity vanishes at every step ($g_n'(u_n) = 0$), the derivation of the order conditions and hence the schemes can be simplified considerably. In particular, higher-order stiffly accurate schemes can be derived with only a few stages (see the next section).

The convergence analysis of exponential Rosenbrock methods is usually carried out in an appropriate framework (strongly continuous semigroup) under regularity assumptions on the solution $u(t)$ (sufficiently smooth) and $g_n(u)$ (sufficiently Fréchet differentiable in a neighborhood of the solution) with uniformly bounded derivatives in some Banach space. For more details, we refer to [26, 32].

3.3 Selected Schemes for Numerical Simulations

First, we discuss some important points for the derivation of exponential Rosenbrock schemes. Clearly, the unknown coefficients $a_{ij}(hJ_n)$ and $b_i(hJ_n)$ have to be determined by solving order conditions. For nonstiff problems, where the Jacobian matrix has a small norm, one can expand those matrix functions using classical Taylor series expansions, leading to nonstiff order conditions and in turn classical exponential schemes (see e.g. [9, 27]). For stiff problems, however, one has to be cautious when analyzing the local error to make sure that error terms do not involve powers of J_n (which has a large norm). Recently, Luan and Ostermann [30, 33] derived a new expansion of the local error which fulfills this requirement and thus derived a new stiff order conditions theory for methods of arbitrary order (both for exponential Runge–Kutta and exponential Rosenbrock methods). As expected, with the same order, the number of order conditions for exponential Rosenbrock methods is significant less than those for exponential Runge–Kutta methods. For example, in Table 1, we display the required 4 conditions for deriving schemes up to order 5 in [32] (note that for exponential Runge–Kutta methods, 16 order conditions are required for deriving schemes of order 5, see [31]).

We note that with these order conditions one can easily derive numerous different schemes of order up to 5. Taking the compromise between efficiency and accuracy into consideration, we seek for the most efficient schemes for our applications. Namely, the following two representative fourth-order schemes are selected.

exprb42 (a fourth-order 2-stage scheme which can be considered as a super-convergent scheme, see [29]):

$$U_{n2} = u_n + \tfrac{3}{4}h\varphi_1(\tfrac{3}{4}hJ_n)F(u_n), \tag{20a}$$

$$u_{n+1} = u_n + h\varphi_1(hJ_n)F(u_n) + h\tfrac{32}{9}\varphi_3(hJ_n)(g_n(U_{n2}) - g_n(u_n)). \tag{20b}$$

Table 1 Stiff order conditions for exponential Rosenbrock methods up to order five. Here Z and K denote arbitrary square matrices and $\psi_{3,i}(z) = \sum_{k=2}^{i-1} a_{ik}(z)\frac{c_k^2}{2!} - c_i^3\varphi_3(c_i z)$

No.	Order condition	Order
1	$\sum_{i=2}^{s} b_i(Z)c_i^2 = 2\varphi_3(Z)$	3
2	$\sum_{i=2}^{s} b_i(Z)c_i^3 = 6\varphi_4(Z)$	4
3	$\sum_{i=2}^{s} b_i(Z)c_i^4 = 24\varphi_5(Z)$	5
4	$\sum_{i=2}^{s} b_i(Z)c_i K \psi_{3,i}(Z) = 0$	5

pexprb43 (a fourth-order 3-stage scheme, which can be implemented in parallel, see [34]):

$$U_{n2} = u_n + \tfrac{1}{2}h\varphi_1(\tfrac{1}{2}hJ_n)F(u_n), \tag{21a}$$

$$U_{n3} = u_n + h\varphi_1(hJ_n)F(u_n), \tag{21b}$$

$$u_{n+1} = u_n + h\varphi_1(hJ_n)F(u_n) + h\varphi_3(hJ_n)(16D_{n2} - 2D_{n3})$$
$$+ h\varphi_4(hJ_n)(-48D_{n2} + 12D_{n3}). \tag{21c}$$

Note that the vectors D_{n2} and D_{n3} in (21) are given by (17c), i.e., $D_{n2} = g_n(U_{n2}) - g_n(u_n)$ and $D_{n3} = g_n(U_{n3}) - g_n(u_n)$.

4   Implementation

In this section, we present the implementation of exponential Rosenbrock methods for the new formulation (5) of the system of coupled oscillators. First, we discuss on the computation of the matrix square root Ω needed for the reformulation. We then briefly review some state-of-the-art algorithms for implementing exponential Rosenbrock methods and introduce a new routine which is an improved version of one of these algorithms (proposed very recently in [35]) for achieving more efficiently. Finally, we specifically discuss applying this routine for implementing the selected schemes exprb42 and pexprb43.

4.1   Computation of the Matrix Square Root $\Omega = \sqrt{A}$

For the computation of $\Omega = \sqrt{A}$ used in (5), we follow our approach in [40]. Specifically, we use the Schur decomposition for moderate systems. For large systems, the Newton square root iteration (see [20]) is employed in order to avoid an explicit precomputation of Ω. Namely, one can use the following simplified iteration method for approximating the solution of the equation $\Omega^2 = A$:

(i) choose $\Omega_0 = A$ ($k = 0$),
(ii) update $\Omega_{k+1} = \tfrac{1}{2}(\Omega_k + \Omega_k^{-1}A)$.

This method offers unconditional quadratic convergence with much less cost compared to the Schur decomposition. We note that Ω^{-1} can be computed efficiently using a Cholesky decomposition since Ω is symmetric and positive definite and it is given by $\Omega^{-1} = S^{-1}S^{-T}$, where S is an upper triangular matrix with real and positive diagonal entries. For more details, we refer to [20, 40].

With Ω at hand, one can easily compute the Jacobian J_n as in (11) and $F(u), G_n(u)$ as in (10). As the next step, we discuss the implementation of the exponential Rosenbrock schemes.

4.2 Implementation of Exponential Rosenbrock Methods

In view of the exponential Rosenbrock schemes in Sect. 3, each stage requires the evaluation of a linear combination of φ-functions acting on certain vectors $v_0, \ldots, v_p$

$$\varphi_0(M)v_0 + \varphi_1(M)v_1 + \varphi_2(M)v_2 + \cdots + \varphi_p(M)v_p, \tag{22}$$

where the matrix M here could be hJ_n or $c_i hJ_n$. Starting from a seminal contribution by Hochbruck and Lubich [21] (which they analyzed Krylov subspace methods for efficiently computing the action of a matrix exponential (with a large norm) on some vector), many more efficient techniques have been proposed. A large portion of these developments is concerned with computing the expression (22). For example, we mention some of the state-of-the-art algorithms: expmv proposed by Al-Mohy and Higham in [1] (using a truncated standard Taylor series expansion), phipm proposed by Niessen and Wright in [41] (using adaptive Krylov subspace methods), and expleja proposed by Caliari et al. in [5, 6] (using Leja interpolation). With respect to computational time, it turns out that phipm offers an advantage. This algorithm utilizes an adaptive time-stepping method to evaluate (22) using only one matrix function (see Sect. 4.2.1 below). This task is carried out in a lower dimensional Krylov subspace using standard Krylov subspace projection methods i.e. the Arnoldi iteration. Moreover, the dimension of Krylov subspaces and the number of substeps are also chosen adaptivity for improving efficiency.

Recently, the phipm routine was modified by Gaudreault and Pudykiewicz in [13] (Algorithm 2) by using the incomplete orthogonalization method (IOM) within the Arnoldi iteration and by adjusting the two crucial initial parameters for starting the Krylov adaptivity. This results in the new routine called phipm/IOM2. It is shown in [13] that this algorithm reduces computational time significantly compared to phipm when integrating the shallow water equations on the sphere.

Very recently, the authors of [35] further improved phipm/IOM2 which resulted in a more efficient routine named as phipm_simul_iom2. For the reader's convenience, we present the idea of the adaptive time-stepping method (originally proposed in [41]) for evaluating (22) and introduce some new features of the new routine phipm_simul_iom2.

4.2.1 Computing of Linear φ-Combinations Based on Time-Stepping

It was observed that the following linear ODE

$$u'(t) = Mu(t) + v_1 + tv_2 + \cdots + \frac{t^{p-1}}{(p-1)!} v_p, \quad u(0) = v_0, \tag{23}$$

defined on the interval $[0, 1]$ has the exact solution at $t = 1$, $u(1)$ to be the expression (22). The time-stepping technique approximates $u(1)$ by discretizing $[0, 1]$ into subintervals $0 = t_0 < t_1 < \cdots < t_k < t_{k+1} = t_k + \tau_k < \cdots < t_K = 1$ with a substepsize sequence τ_k ($k = 0, 1, \ldots, K - 1$) and using the following relation between $u(t_{k+1})$ and its previous solution $u(t_k)$:

$$u(t_{k+1}) = \varphi_0(\tau_k M) u(t_k) + \sum_{i=1}^{p} \tau_k^i \varphi_i(\tau_k M) \sum_{j=0}^{p-i} \frac{t_k^j}{j!} v_{i+j}. \tag{24}$$

Using the recursion relation (19) and (24) can be simplified as

$$u(t_{k+1}) = \tau_k^p \varphi_p(\tau_k M) w_p + \sum_{j=0}^{p-i} \frac{t_k^j}{j!} w_j, \tag{25}$$

where the vectors w_j satisfy the recurrence relation

$$w_0 = u(t_k), \quad w_j = Mw_{j-1} + \sum_{\ell=0}^{p-j} \frac{t_k^\ell}{\ell!} v_{j+\ell}, \quad j = 1, \ldots, p. \tag{26}$$

Equation (25) implies that evaluating $u(t_K) = u(1)$ i.e. the expression (22) requires only one matrix function $\varphi_p(\tau_k A) w_p$ in each substep instead of $(p + 1)$ matrix-vector multiplications. As $0 < \tau_k < 1$, this task can be carried out in a Krylov subspace of lower dimension m_k, and in each substep only one Krylov projection is needed. With a reasonable number of substeps K, it is thus expected that the total computational cost of $O(m_1^2) + \cdots + O(m_K^2)$ for approximating $\varphi_p(\tau_k M) w_p$ is less than that of $O(m^2)$ for approximating $\varphi_p(M) v$ in a Krylov subspace of dimension m. If K is too large (e.g. when the spectrum of M is very large), this might be not true. This case, however, is handed by using the adaptive Krylov algorithm in [41] allowing to adjust both the dimension m and the step sizes τ_k adaptivity. This explains the computational advance of this approach compared to standard Krylov algorithms.

4.2.2 New Routine `phipm_simul_iom2` [35]

First, we note that the resulting routine `phipm_simul_iom2` optimizes computational aspects of `phipm/IOM2` corresponding to the following two specific changes:

1. Unlike (22), where each of the φ_k functions is evaluated at the same argument M, the internal stages of exponential Rosenbrock schemes require evaluating the φ functions at fractions of the matrix M:

$$w_k = \sum_{l=1}^{p} \varphi_l(c_k M)v_l, \quad k = 2, \ldots, s, \tag{27}$$

where now the node values $c_2, \ldots, c_s$ are scaling factors used for each v_k output. To optimize this evaluation, `phipm_simul_iom2` computes all w_k outputs in (27) simultaneously, instead of computing only one at a time. This is accomplished by first requiring that the entire array $c_2, \ldots, c_s$ as an input to the function. Within the substepping process (24), each value c_j is aligned with a substep-size τ_k. The solution vector is stored at each of these moments and on output the full set $\{w_k\}_{k=1}^s$ is returned. Note that this approach is similar but differs from [48] that it guarantees no loss of solution accuracy since it explicitly stops at each c_k instead of using interpolation to compute w_k as in [48].
2. In view of the higher-order exponential Rosenbrock schemes (see also from Sect. 3.3), it is realized that they usually use a subset of the φ_l functions. Therefore, multiple vectors in (27) will be zero. In this case, `phipm_simul_iom2` will check whether $w_{j-1} \neq 0$ (within the recursion (26)) before computing the matrix-vector product $M w_{j-1}$. While matrix-vector products require $O(n^2)$ work, checking $u \neq 0$ requires only $O(n)$. This can result in significant savings for large n.

4.2.3 Implementation of `exprb42` and `pexprb43`

Taking a closer look at the structures of the two selected exponential Rosenbrock schemes `exprb42` and `pexprb43`, we now make use of `phipm_simul_iom2` for implementing these schemes. For simplicity, let us denote $M = hJ_n$ and $v = hF(u_n)$.

Implementation of `exprb42`: Due to the structure of `exprb42` given in (20), one needs two calls to `phipm_simul_iom2`:

1. Evaluate $y_1 = \varphi_1(\frac{3}{4}M)w_1$ with $w_1 = \frac{3}{4}v$ (so $w_0 = 0$) to get $U_{n2} = u_n + y_1$,
2. Evaluate $w = \varphi_1(M)v_1 + \varphi_3(M)v_3$ (i.e. $v_0 = v_2 = 0$) with $v_1 = v$, $v_3 = \frac{32}{9}hD_{n2}$ to get $u_{n+1} = u_n + w$.

Implementation of `pexprb43`: Although `pexprb43` is a 3-stage scheme, its special structure (21) allows to use only two calls to `phipm_simul_iom2`:

1. Evaluate both terms $y_1 = \varphi_1(\frac{1}{2}M)v$ and $z_1 = \varphi_1(M)v$ simultaneously to get the two stages $U_{n2} = u_n + \frac{1}{2}y_1$ and $U_{n3} = u_n + z_1$,
2. Evaluate $w = \varphi_3(M)v_3 + \varphi_4(M)v_4$ (i.e. $v_0 = v_1 = v_2 = 0$) with $v_3 = h(16D_{n2} - 2D_{n3})$, $v_4 = h(-48D_{n2} + 12D_{n3})$ to get $u_{n+1} = U_{n3} + w$.

5 Numerical Examples

In this section we present several numerical examples to study the behavior of the presented exponential Rosenbrock-type methods, in particular the fourth-order scheme `exprb42` using two stages and the fourth-order `pexprb43` scheme using three stages implemented in parallel.

In particular, we focus on relevant aspects in the realm of visual computing, like stability and energy conservation, large stiffness, and high fidelity and visual accuracy. A tabular summary of the models that are used throughout this section can be found in Table 2. Furthermore, our simulation includes important aspects like elastic collisions and nonelastic deformations. The presented exponential Rosenbrock-type methods are evaluated against classical and state-of-the-art methods used in visual computing, in particular against the implicit-explicit variational (IMEX) integrator (cf. [44, 45]), the standard fourth-order Runge–Kutta method (see [28, 43]), and the implicit BDF-1 integrator (see [11]). All simulation results visualized here have been computed using a machine with an Intel(R) Xeon E5 3.5 GHz and 32 GB DDR-RAM. For each simulation scenario the largest possible time step size is used which still leads to a desired visually plausible result.

5.1 Simulation of Deformable Bodies

In order to illustrate the accurate energy preservation of the presented exponential Rosenbrock-type methods, we set up an undamped scene of an oscillating coil spring, which is modeled as a deformable body composed of tetrahedra, in particular of 8000 vertices corresponding to $N = 24\,000$ equations of motion, which are derived from a system of coupled oscillators with uniform spring stiffness of $k = 10^6$. Since the coil spring is exposed to an external forces field, it starts to oscillate as illustrated in Fig. 1. It can be seen that the top of the coil spring returns to its initial height periodically during the simulation which can be seen as an indicator for energy conservation. In fact when using the exponential Rosenbrock-type methods `exprb42` and `pexprb43` we observe that the discrete energy is only slightly oscillating around the real energy without increasing oscillations over time. In contrast, the standard fourth order Runge–Kutta method respectively the BDF-1

Table 2 Overview of the test cases used for the numerical experiments. Their complexity N (i.e. the number of the resulting equations of motion), the simulated time, and respective running times for the exponential Rosenbrock-type methods `exprb42` and `pexprb43`, the implicit-explicit variational integrator (V-IMEX), the standard fourth order Runge–Kutta method (RK4), and the BDF-1 integrator are shown

No.	Model	N	Sim. time	`exprb42`	`pexprb43`	V-IMEX	RK 4	BDF-1
1	Coil Spring	24k	60 s	55 s	47 s	12 min	46 min	62 min
2	Brushing	90k	15 s	52 s	51 s	11 min	53 min	72 min
3	Crash test (moderate)	360k	2 s	44 s	44 s	9 min	47 min	58 min
4	Crash test (fast)	360k	2 s	47 s	46 s	9 min	47 min	59 min

Fig. 1 Simulation of an oscillating coil spring

integrator generate significant numerical viscosity leading to a loss of energy around 22% respectively 40% after 60 s of simulated time.

The exponential Rosenbrock-type methods `exprb42` and `pexprb43` show their advantageous behavior since these methods can be applied with orders of magnitude larger time steps compared to the other integrators. Even with a step size of $h = 0.05$ the relative error is still below 2% for `exprb42` and about a single percent for `pexprb43`.[1] From a point of view of computation time, we achieve a speed up of a factor of around thirteen using `exprb42` and of over fifteen using `pexprb43` compared to the second best method, the variational IMEX integrator as illustrated in Table 2. Compared to the other methods, the exponential Rosenbrock-type methods allow for accurate simulations in real-time.

5.2 Simulation of Fibers Including Elastic Collisions

Fibers are canonical examples for complex interacting systems. According to the work of Michels et al. (see [39]), we set up a toothbrush composed of individual bristles. Each bristle consists of coupled oscillators that are connected in such a way that the fiber axis is enveloped by a chain of cuboid elements. For preventing a volumetric collapse during the simulation, additional diagonal springs are used. The toothbrush consists of 1500 bristles, each of 20 particles leading to 90 000 equations of motion. We make use of additional repulsive springs in order to prevent from interpenetrations.[2] Since the approach allows for the direct use of realistic parameters in order to set up the stiffness values in the system of coupled oscillators, we employ a Young's modulus of $3.2 \cdot 10^6 \, \mathrm{Ncm}^{-2}$, a torsional modulus of $10^5 \, \mathrm{Ncm}^{-2}$, and segment thicknesses of 0.12 mm.

---

[1] We estimated the error after 60 s of simulated time based on the accumulated Euclidean distances of the individual particles in the position space compared to ground truth values which are computed with a sufficiently small step size.

[2] In order to detect collisions efficiently, we make use of a standard bounding volume hierarchy.

Fig. 2 Simulation of a brush cleaning a bronze-colored paperweight

Fig. 3 Simulations of two frontal nonelastic crash scenarios: a car with moderate velocity (top) and high velocity (bottom)

We simulate 15 s of a toothbrush cleaning a paperweight illustrated in Fig. 2. This simulation can be carried out almost in real-time which is not possible with the use of classical methods as illustrated in Table 2.

5.3 Crash Test Simulation Including Nonelastic Deformations

As a very complex example with relevance in the context of special effects, we simulate a frontal crash of a car into a wall as illustrated in Fig. 3. The mesh of the car and its interior is composed of 120 000 vertices leading to 360 000 equations of motion. The global motion (i.e. the rebound of the car) is computed by treating the car as a rigid body. Using an appropriate bounding box, this can be easily carried out in real-time. The deformation is then computed using a system of coupled oscillators with structural stiffness values of $k = 10^4$ and bending stiffness values of $k/100$. If the deformation reaches a defined threshold, the rest lengths of the corresponding springs are corrected in a way, that they do not elastically return to their initial shape. Using the exponential Rosenbrock-type methods, the whole simulation can be carried out at interactive frame rates. Such an efficient computation can not be achieved with established methods as illustrated in Table 2.

6 Conclusion

We introduced the class of *explicit exponential Rosenbrock methods* for the time integration of large systems of nonlinear differential equations. In particular, the exponential Rosenbrock-type fourth-order schemes `exprb42` using two stages and `pexprb43` using three stages were discussed and their implementation were addressed. In order to study their behavior, a broad spectrum of numerical examples was computed. In this regard, the simulation of deformable bodies, fibers including elastic collisions, and crash scenarios including nonelastic deformations was addressed focusing on relevant aspects in the realm of visual computing, like stability and energy conservation, large stiffness values, and high fidelity and visual accuracy. An evaluation against classical and state-of-the-art methods was presented demonstrating their superior performance with respect to the simulation of large systems of stiff differential equations.

Acknowledgments The first author has been partially supported by NSF grant DMS–2012022. The second author has been partially supported by King Abdullah University of Science and Technology (KAUST baseline funding).

References

1. A.H. Al-Mohy, N.J. Higham, Computing the action of the matrix exponential, with an application to exponential integrators. SIAM J. Sci. Comput. **33**, 488–511 (2011)
2. U. Ascher, S. Ruuth, B. Wetton, Implicit-explicit methods for time-dependent PDEs. SIAM J. Numer. Anal. **32**(3), 797–823 (1997)
3. D. Baraff, A. Witkin, Large steps in cloth simulation, in *ACM Transactions on Graphics, SIGGRAPH'98* (ACM, New York, 1998), pp. 43–54
4. M. Bergou, M. Wardetzky, S. Robinson, B. Audoly, E. Grinspun, Discrete elastic rods. ACM Trans. Graph. **27**(3), 63:1–63:12 (2008)
5. M. Caliari, A. Ostermann, Implementation of exponential Rosenbrock-type integrators. Appl. Numer. Math. **59**(3–4), 568–581 (2009)
6. M. Caliari, P. Kandolf, A. Ostermann, S. Rainer, The Leja method revisited: backward error analysis for the matrix exponential. SIAM J. Sci. Comput. **38**(3), A1639–A1661 (2016)
7. W.L. Chao, J. Solomon, D. Michels, F. Sha, Exponential integration for Hamiltonian Monte Carlo, in *Proceedings of the 32nd International Conference on Machine Learning*, ed. by F. Bach, D. Blei. Proceedings of Machine Learning Research, vol. 37, pp. 1142–1151 (PMLR, Lille, 2015)
8. Y.J. Chen, U.M. Ascher, D.K. Pai, Exponential Rosenbrock-Euler integrators for elastodynamic simulation. IEEE Trans. Visual. Comput. Graph. **24**(10), 2702–2713 (2018)
9. S.M. Cox, P.C. Matthews, Exponential time differencing for stiff systems. J. Comput. Phys. **176**(2), 430–455 (2002)
10. C. Curtiss, J.O. Hirschfelder, Integration of stiff equations. Proc. Natl. Acad. Sci. **38**(3), 235–243 (1952)
11. C.F. Curtiss, J.O. Hirschfelder, Integration of stiff equations. Proc. Natl. Acad. Sci. USA **38**(3), 235–243 (1952)

12. B. Eberhardt, O. Etzmuß, M. Hauth, Implicit-explicit schemes for fast animation with particle systems, in *Proceedings of the 11th Eurographics Workshop on Computer Animation and Simulation (EGCAS)* (Springer, Berlin, 2000), pp. 137–151
13. S. Gaudreault, J. Pudykiewicz, An efficient exponential time integration method for the numerical solution of the shallow water equations on the sphere. J. Comput. Phys. **322**, 827–848 (2016)
14. C. Gear, *Numerical Initial Value Problems in Ordinary Differential Equations* (Prentice–Hall, Englewood Cliffs, 1971)
15. S. Geiger, G. Lord, A. Tambue, Exponential time integrators for stochastic partial differential equations in 3D reservoir simulation. Comput. Geosci. **16**(2), 323–334 (2012)
16. R. Goldenthal, D. Harmon, R. Fattal, M. Bercovier, E. Grinspun, Efficient simulation of inextensible cloth, in *ACM Transactions on Graphics, SIGGRAPH'07* (2007)
17. M.A. Gondal, Exponential Rosenbrock integrators for option pricing. J. Comput. Appl. Math. **234**(4), 1153–1160 (2010)
18. E. Hairer, G. Wanner, *Solving Ordinary Differential Equations II: Stiff and Differential-Algebraic Problems* (Springer, New York, 1996)
19. M. Hauth, O. Etzmuss, A high performance solver for the animation of deformable objects using advanced numerical methods. Comput. Graph. Forum **20**, 319–328 (2001)
20. N.J. Higham, *Functions of Matrices: Theory and Computation* (SIAM, Philadelphia, 2008)
21. M. Hochbruck, C. Lubich, On Krylov subspace approximations to the matrix exponential operator. SIAM J. Numer. Anal. **34**, 1911–1925 (1997)
22. M. Hochbruck, A. Ostermann, Explicit exponential Runge–Kutta methods for semilinear parabolic problems. SIAM J. Numer. Anal. **43**, 1069–1090 (2005)
23. M. Hochbruck, A. Ostermann, Explicit integrators of Rosenbrock-type. Oberwolfach Rep. **3**, 1107–1110 (2006)
24. M. Hochbruck, A. Ostermann, Exponential integrators. Acta Numer. **19**, 209–286 (2010)
25. M. Hochbruck, C. Lubich, H. Selhofer, Exponential integrators for large systems of differential equations. SIAM J. Sci. Comput. **19**, 1552–1574 (1998)
26. M. Hochbruck, A. Ostermann, J. Schweitzer, Exponential Rosenbrock-type methods. SIAM J. Numer. Anal. **47**, 786–803 (2009)
27. S. Krogstad, Generalized integrating factor methods for stiff PDEs. J. Comput. Phys. **203**(1), 72–88 (2005)
28. M.W. Kutta, Beitrag zur näherungsweisen Integration totaler Differentialgleichungen. Z. Math. Phys. **46**, 435–453 (1901)
29. V.T. Luan, Fourth-order two-stage explicit exponential integrators for time-dependent PDEs. Appl. Numer. Math. **112**, 91–103 (2017)
30. V.T. Luan, A. Ostermann, Exponential B-series: the stiff case. SIAM J. Numer. Anal. **51**, 3431–3445 (2013)
31. V.T. Luan, A. Ostermann, Explicit exponential Runge–Kutta methods of high order for parabolic problems. J. Comput. Appl. Math. **256**, 168–179 (2014)
32. V.T. Luan, A. Ostermann, Exponential Rosenbrock methods of order five–construction, analysis and numerical comparisons. J. Comput. Appl. Math. **255**, 417–431 (2014)
33. V.T. Luan, A. Ostermann, Stiff order conditions for exponential Runge–Kutta methods of order five, in *Modeling, Simulation and Optimization of Complex Processes - HPSC 2012*, H.B. et al. (ed.) (Springer, Berlin, 2014), pp. 133–143
34. V.T. Luan, A. Ostermann, Parallel exponential Rosenbrock methods. Comput. Math. Appl. **71**, 1137–1150 (2016)
35. V.T. Luan, J.A. Pudykiewicz, D.R. Reynolds, Further development of the efficient and accurate time integration schemes for meteorological models J. Comput. Sci. **376**, 817–837 (2018)
36. D.L. Michels, M. Desbrun, A semi-analytical approach to molecular dynamics. J. Comput. Phys. **303**, 336–354 (2015)
37. D.L. Michels, J.P.T. Mueller, Discrete computational mechanics for stiff phenomena, in *SIGGRAPH ASIA 2016 Courses* (2016), pp. 13:1–13:9

38. D.L. Michels, G.A. Sobottka, A.G. Weber, Exponential integrators for stiff elastodynamic problems. ACM Trans. Graph. **33**(1), 7:1–7:20 (2014)
39. D.L. Michels, J.P.T. Mueller, G.A. Sobottka, A physically based approach to the accurate simulation of stiff fibers and stiff fiber meshes. Comput. Graph. **53B**, 136–146 (2015)
40. D.L. Michels, V.T. Luan, M. Tokman, A stiffly accurate integrator for elastodynamic problems. ACM Trans. Graph. **36**(4), 116 (2017)
41. J. Niesen, W.M. Wright, Algorithm 919: a Krylov subspace algorithm for evaluating the φ-functions appearing in exponential integrators. ACM Trans. Math. Softw. **38**, 3 (2012)
42. D.A. Pope, An exponential method of numerical integration of ordinary differential equations. Commun. ACM **6**, 491–493 (1963)
43. C.D. Runge, Über die numerische Auflösung von Differentialgleichungen. Math. Ann. **46**, 167–178 (1895)
44. A. Stern, M. Desbrun, Discrete geometric mechanics for variational time integrators, in *SIGGRAPH 2006 Courses* (2006), pp. 75–80
45. A. Stern, E. Grinspun, Implicit-explicit variational integration of highly oscillatory problems. Multiscale Model. Simul. **7**, 1779–1794 (2009)
46. A. Tambue, I. Berre, J.M. Nordbotten, Efficient simulation of geothermal processes in heterogeneous porous media based on the exponential Rosenbrock–Euler and Rosenbrock-type methods. Adv. Water Resour. **53**, 250–262 (2013)
47. D. Terzopoulos, J. Platt, A. Barr, K. Fleischer, Elastically deformable models. ACM Trans. Graph. **21**, 205–214 (1987)
48. M. Tokman, J. Loffeld, P. Tranquilli, New adaptive exponential propagation iterative methods of Runge-Kutta type. SIAM J. Sci. Comput. **34**, A2650–A2669 (2012)
49. H. Zhuang, I. Kang, X. Wang, J.H. Lin, C.K. Cheng, Dynamic analysis of power delivery network with nonlinear components using matrix exponential method, in *2015 IEEE Symposium on Electromagnetic Compatibility and Signal Integrity* (IEEE, Piscataway, 2015), pp. 248–252

W-Methods and Approximate Matrix Factorization for Parabolic PDEs with Mixed Derivative Terms

Severiano González-Pinto and Domingo Hernández-Abreu

1 ADI and W-Methods

This chapter deals with the time integration of parabolic partial differential equations (PDEs) with mixed derivative terms discretized by means of the method of lines (MoL). On an m-dimensional box, which for ease of presentation we take $\Omega = (0, 1)^m \subset \mathbb{R}^m$, and for $t > 0$ we consider the PDE problem

$$
\begin{aligned}
\partial_t u &= \sum_{i,j=1}^{m} \alpha_{i,j}(t, \mathbf{x})\, \partial_{x_i x_j}^2 u + \sum_{j=1}^{m} \eta_j(t, \mathbf{x})\partial_{x_j} u + g(t, \mathbf{x}, u) \\
&\quad \mathbf{x} = (x_1, \ldots, x_m)^\top \in \Omega, \quad t \in (0, T], \\
u(t, \mathbf{x}) &= \beta(t, \mathbf{x}), \quad (t, \mathbf{x}) \in (0, T] \times \partial\Omega, \\
u(0, \mathbf{x}) &= u_0(\mathbf{x}), \quad \mathbf{x} \in \Omega,
\end{aligned}
\tag{1}
$$

where $\partial\Omega$ denotes the boundary of Ω ($\bar{\Omega} = \Omega \cup \partial\Omega$), $g(t, \mathbf{x}, u)$ stands for the reaction terms, $\eta_j(t, \mathbf{x})$ corresponds to advection terms on each space variable and the diffusion terms are those corresponding to the coefficient matrix $\mathscr{A} = (\alpha_{i,j}(t, \mathbf{x}))_{i,j=1}^{m}$, which is assumed to be symmetric and positive definite for each $(t, \mathbf{x}) \in [0, T] \times \bar{\Omega}$. In the sequel $\mathscr{A} > 0$ indicates that $\mathscr{A}$ is a positive definite matrix. The PDE problem is provided with an initial condition and Dirichlet boundary conditions.

S. González-Pinto (✉) · D. Hernández-Abreu
Departamento de Análisis Matemático, Universidad de La Laguna, La Laguna, Spain
e-mail: spinto@ull.edu.es; dhabreu@ull.edu.es

© The Author(s), under exclusive license to Springer Nature Switzerland AG 2021
T. Jax et al. (eds.), *Rosenbrock—Wanner—Type Methods*, Mathematics Online
First Collections, https://doi.org/10.1007/978-3-030-76810-2_4

With a space discretization of (1) by means of Finite Differences (or Finite Volumes) large systems of Ordinary Differential Equations (ODEs) arise

$$\dot{U} = F(t, U), \quad U(0) = U_0, \quad t \in [0, T], \tag{2}$$

where $U(t)$ is a real vector approximating the solution values at grid points, and $F(t, U)$ collects the terms of the spatial discretization, reaction terms, and the contribution of inhomogeneous boundary conditions. Inspired by the Alternating Direction Implicit (ADI) approach [2, 23], the function $F(t, U)$ is typically split as

$$F(t, U) = \sum_{j=0}^{m} F_j(t, U), \tag{3}$$

where for each $j = 1, \ldots, m$, $F_j(t, U)$ contains the terms corresponding to space derivatives with respect to x_j (including boundary conditions). Here, it is assumed that $F_0(t, U)$ includes the terms corresponding to the mixed derivatives and their respective boundary conditions, as well as the discretization of the reaction terms, which are assumed to be non-stiff or mildly stiff.

Alternating Direction Implicit schemes, in the absence of mixed derivatives, were proposed by Peaceman, Rachford, and Douglas (see [23] and [2]) in order to reduce the computational cost in the solution of the arising linear systems to the level of one dimensional problems (with matrices having a banded structure with small bandwidths). Craig and Sneyd [1] then came up with a second order scheme for parabolic problems with mixed derivatives. More recently, other ADI schemes of order two have become popular for the time integration of parabolic problems with mixed derivatives, in particular in the context of applications in financial mathematics. Examples of such schemes are the *Hundsdorfer–Verwer* (HV) method [16, 17, 19]

$$\begin{aligned}
Y_0 &= U_n + \tau F(t_n, U_n), \\
Y_j &= Y_{j-1} + \theta\tau\big(F_j(t_{n+1}, Y_j) - F_j(t_n, U_n)\big), \quad j = 1, \ldots, m, \\
\tilde{Y}_0 &= Y_0 + \mu\tau\big(F(t_{n+1}, Y_m) - F(t_n, U_n)\big), \\
\tilde{Y}_j &= \tilde{Y}_{j-1} + \theta\tau\big(F_j(t_{n+1}, \tilde{Y}_j) - F_j(t_{n+1}, Y_m)\big), \quad j = 1, \ldots, m, \\
U_{n+1} &= \tilde{Y}_m,
\end{aligned} \tag{4}$$

with $\mu = 1/2$ to get classical order two (and $\theta > 0$ for stability), and the *modified Craig-Sneyd* (MCS) scheme [19, 20]

$$
\begin{aligned}
Y_0 &= U_n + \tau F(t_n, U_n), \\
Y_j &= Y_{j-1} + \theta\tau\big(F_j(t_{n+1}, Y_j) - F_j(t_n, U_n)\big), \quad j = 1, \ldots, m, \\
\widehat{Y}_0 &= Y_0 + \sigma\tau\big(F_0(t_{n+1}, Y_m) - F_0(t_n, U_n)\big), \\
\widetilde{Y}_0 &= \widehat{Y}_0 + \mu\tau\big(F(t_{n+1}, Y_m) - F(t_n, U_n)\big), \\
\widetilde{Y}_j &= \widetilde{Y}_{j-1} + \theta\tau\big(F_j(t_{n+1}, \widetilde{Y}_j) - F_j(t_n, U_n)\big), \quad j = 1, \ldots, m, \\
U_{n+1} &= \widetilde{Y}_m,
\end{aligned}
\tag{5}
$$

with parameters $\sigma = \theta$ and $\mu = \frac{1}{2} - \theta$ to get order two and $\theta > 0$. The original second order Craig–Sneyd scheme [1] is obtained from (5) when $\mu = 0$ and $\sigma = \theta = \frac{1}{2}$. Above, $\tau > 0$ stands for the time stepsize to advance from (t_n, U_n) to (t_{n+1}, U_{n+1}).

Both schemes (4) and (5) are extensions of the Douglas scheme [2], which is obtained by considering just the first two lines of either methods, but this latter method is only order one when $F_0 \neq 0$. The HV scheme has been recently considered together with space discretizations of order 4 in [3] and applied to stochastic volatility models in financial option pricing in [4]. Compact schemes of order 4 in space based on both the MCS and the HV schemes have been also recently treated in [13, 14].

In this chapter our focus is on W-methods [27] and [11, Section IV.7], which avoid the solution of nonlinear equations and only require an approximate solution of linear systems with matrix $I - \theta\tau W$, where I is the identity, θ is a real parameter, τ the time step size, and W is an approximation to the Jacobian matrix of the ODE. W-methods do not require the solution of nonlinear systems and they allow the use of non-exact approximations for the Jacobian of the vector field, providing both a high classical order and good stability properties.

Considering time t as an independent variable, and augmenting (2) with $\dot{t} = 1$ yields for $y = (t, U)^\top$ an autonomous system

$$
\dot{y} = f(y), \quad y(0) = y_0, \quad t \in [0, T].
\tag{6}
$$

The splitting (3) leads to a splitting of the form

$$
\dot{y}(t) = f(y) = \sum_{j=0}^{m} f_j(y) = \begin{pmatrix} 1 \\ F_0(t, U) \end{pmatrix} + \sum_{j=1}^{m} \begin{pmatrix} 0 \\ F_j(t, U) \end{pmatrix},
\tag{7}
$$

for which the corresponding splitting for the full Jacobian is then given by

$$\partial_y f(y_n) = \sum_{j=0}^{m} \partial_y f_j(y_n) = \sum_{j=0}^{m} \begin{pmatrix} 0 & 0 \\ \partial_t F_j(t_n, U_n) & \partial_U F_j(t_n, U_n) \end{pmatrix}. \tag{8}$$

Now, for the autonomous problem (6), let y_n be a numerical approximation to $y(t)$ at t_n. Then, with a stepsize $\tau > 0$, the numerical approximation y_{n+1} provided by a s-stage W-method at $t_{n+1} = t_n + \tau$ is defined by

$$(I - \theta \tau W)\tilde{K}_i = \tau f \left(y_n + \sum_{j=1}^{i-1} a_{i,j} \tilde{K}_j \right) + \sum_{j=1}^{i-1} \ell_{i,j} \tilde{K}_j, \quad i = 1, 2, \dots, s,$$

$$y_{n+1} = y_n + \sum_{i=1}^{s} b_i \tilde{K}_i. \tag{9}$$

The matrix W is arbitrary, but it is intended to approximate $\partial_y f(y_n)$. For $W = \partial_y f(y_n)$ we obtain the underlying ROW or Rosenbrock method. It is characterized by the coefficients (A, L, b, θ), where $A = (a_{i,j})_{j<i}$, $L = (\ell_{i,j})_{j<i}$ and $b = (b_i)_i$. The coefficient matrix $A(I - L)^{-1}$ and the weight vector $b^T(I - L)^{-1}$ define the underlying explicit Runge-Kutta method associated to the W-method (see, e.g., [8]).

In the literature, several options for the selection of the matrix W have been considered. Some methods up to order of consistency four under the assumption

$$W - \partial_y f(y_n) = \mathcal{O}(\tau), \qquad \tau \to 0, \tag{10}$$

have been introduced in [5, 8, 22, 25]. A more general situation where the commutator satisfies

$$[W, \partial_y f(y_n)] := W \partial_y f(y_n) - \partial_y f(y_n)W = \mathcal{O}(\tau), \qquad \tau \to 0, \tag{11}$$

was studied in [7] and some families of third order methods under such assumption were presented. The construction of efficient W-methods of order ≥ 3 in the general setting $W - \partial_y f(y_n) = \mathcal{O}(1)$ is a demanding task due to the high number of order conditions to be satisfied, see e.g. [11, 21]. In [24] some W-methods of order four and six stages have been built. It is worth to mention that the assumptions in (10) and (11) are in ODE sense, since negative powers of the space resolutions are present in the Jacobian matrices and in their approximations W.

In this chapter time integrators that can be applied to general problems of the form (1) are considered, although the emphasis is on a stability analysis that gives insight into the linear diffusion problem

$$\partial_t u = \sum_{i,j=1}^{m} \alpha_{i,j} \partial^2_{x_i x_j} u \tag{12}$$

with homogeneous Dirichlet boundary conditions and a constant positive definite coefficient matrix $\mathscr{A} = (\alpha_{i,j})_{i,j=1}^m$ so that the right-hand side represents an elliptic operator. A standard central finite difference discretization yields $\dot{U} = \mathscr{M}U$, where

$$
\begin{aligned}
\mathscr{M} = {} & \sum_{i=1}^m \alpha_{i,i} (I_{n_{x_m}} \otimes \ldots \otimes D_{x_i x_i} \otimes \ldots \otimes I_{n_{x_1}}) \\
& + 2 \sum_{1 \le i < j \le m} \alpha_{i,j} (I_{n_{x_m}} \otimes \ldots \otimes D_{x_j} \otimes \ldots \otimes D_{x_i} \otimes \ldots \otimes I_{n_{x_1}}),
\end{aligned}
\tag{13}
$$

where I_p denotes the identity matrix of dimension p and $D_{x_i x_i}$ and D_{x_i} are banded differentiation matrices approaching the second and first order spatial derivatives, respectively, placed in the $(m - i + 1)$th position of the tensor product. For the usual second order central discretization $D_{x_i x_i}$ and D_{x_i} are tridiagonal matrices with entries $(1, -2, 1)/\Delta x_i^2$ and $(-1, 0, 1)/(2\Delta x_i)$, respectively, where $\Delta x_i = 1/(n_{x_i} + 1)$ is the spacing in the x_i direction.

The analysis of unconditional stability on linear diffusion problems with constant coefficients for the schemes (4)-(5) in the case of periodic boundary conditions and some general finite difference discretizations for the mixed derivatives was carried out in [19]. From [19, Table 1] we borrow the following Table 1 indicating the values of $\theta \ge \theta_0$ for which the schemes (4) and (5) are unconditionally stable when applied to problems of the form (12). We also point out that the second order Craig–Sneyd scheme (obtained from (5) with $\mu = 0$ and $\sigma = \theta = \frac{1}{2}$) is unconditionally stable whenever $m = 2, 3$, but not for $m \ge 4$.

In Sect. 2 three different options to produce W-methods based on the Approximate Matrix Factorization (AMF), see e.g. [17, 28], for the time integration of (2) and (3) are introduced. These schemes are obtained in terms of the selection of the preconditioner $(I - \theta \tau W)$ in (9) and they slightly differ in computational cost, stability properties and consistency order (in ODE sense). However, the computational costs are quite reasonable, since very few function evaluations and linear systems solves with small bandwidth per integration step are required. In Sect. 3 the unconditional stability for these AMF-type W-methods on linear parabolic problems with mixed derivatives and constant coefficients (12) is analyzed.

Table 1 Values of $\theta \ge \theta_0$ providing unconditional stability for the schemes (4) and (5), respectively, on the problem (12)

m	2	3	4	5	6	7	8	9
HV (4)	0.293	0.402	0.515	0.630	0.745	0.860	0.975	1.091
MCS (5)	0.333	0.462	0.593	0.726	0.860	0.994	1.128	1.262

2   W-Methods Based on AMF-Type Splitting

To solve the linear equations in (9) we set

$$\tilde{K}_i = \begin{pmatrix} \tau \rho_i \\ K_i \end{pmatrix}, \quad \rho_i \in \mathbb{R}, \quad i = 1, \ldots, s. \tag{14}$$

According to (8), the matrix W approximating $\partial_y f(y_n)$ will be required to have the structure

$$W = \begin{pmatrix} 0 & \mathbf{0} \\ \times & \mathbf{X} \end{pmatrix}.$$

Then, from (9) and (7), for each stage $(i = 1, \ldots, s)$ we deduce that

$$(I - \theta \tau W) \begin{pmatrix} \tau \rho_i \\ K_i \end{pmatrix} = \left[\tau \left(\begin{matrix} 1 \\ F(t_n + c_i \tau, U_n + \sum_{j=1}^{i-1} a_{ij} K_j) \end{matrix} \right) + \sum_{j=1}^{i-1} \ell_{ij} \begin{pmatrix} \tau \rho_j \\ K_j \end{pmatrix} \right],$$

$$\begin{pmatrix} t_{n+1} \\ U_{n+1} \end{pmatrix} = \begin{pmatrix} t_n \\ U_n \end{pmatrix} + \sum_{i=1}^{s} b_i \begin{pmatrix} \tau \rho_i \\ K_i \end{pmatrix},$$

$$\tag{15}$$

with

$$\rho = (\rho_i)_{i=1}^{s} = (I - L)^{-1} \mathbf{1}, \quad c = (c_i)_{i=1}^{s} = A\rho. \tag{16}$$

Observe that order of consistency one for W-methods implies $b^T \rho = 1$, hence $t_{n+1} = t_n + \tau$ as expected. Henceforth, we use the following notations to describe the methods

$$A_{n,j} := \partial_U F_j(t_n, U_n), \quad a_{n,j} = \partial_t F_j(t_n, U_n), \quad j = 0, 1, \ldots, m. \tag{17}$$

Next, based on W-methods for the numerical solution of initial value problems in ordinary differential equations (ODEs), three different families of AMF-type methods will be proposed. These families mainly differ in the choice of the W-matrix. For the first family, denoted as AMF-W-methods, the corresponding W-choice is directional (ADI-type) and is an order-zero approximation to the true ODE-Jacobian. The second one, denoted as PDE-W-methods, was introduced in [10] and represents an alternative to produce W-matrices with first order of approximation to the ODE-Jacobian. The third family, denoted as AMFR-W-methods also provides W-matrices with first order of approximation to the ODE-Jacobian, but allows the introduction of a free parameter to improve the stability properties of the methods and it is based on applying linear refinements to the stages of the first family of methods.

2.1 AMF-W Methods

In this case the choice for $(I - \tau\theta W)$ in (15) is based on the Approximate Matrix Factorization, but neglecting in it the Jacobian terms corresponding to the mixed derivatives, i.e.,

$$(I - \theta\tau W) = \prod_{j=1}^{m} \begin{pmatrix} 1 & 0 \\ -\theta\tau a_{n,j} & (I - \tau\theta A_{n,j}) \end{pmatrix}. \tag{18}$$

Here, and in the rest of the chapter, the product of matrices is defined as $\prod_{j=1}^{m} M_j = M_m \ldots M_1$.

By performing the calculations in (15) it is not difficult to check that the stages are computed one after the other (for $i = 1, \ldots, s$) by the formula

$$K_i^{(0)} = \tau F(t_n + c_i\tau, U_n + \sum_{j=1}^{i-1} a_{ij}K_j) + \sum_{j=1}^{i-1} \ell_{ij}K_j$$
$$(I - \theta\tau A_{n,j})K_i^{(j)} = K_i^{(j-1)} + \theta\rho_i\tau^2 a_{n,j}, \quad (j = 1, \ldots, m) \tag{19}$$
$$K_i = K_i^{(m)}.$$

The numerical solution after one step is then given by

$$U_{n+1} = U_n + \sum_{i=1}^{s} b_i K_i. \tag{20}$$

2.2 PDE-W Methods

PDE-W-methods were introduced in [10, Section 5] and they represent a modification of the AMF-W method (19) so that (10) is satisfied. Observe that $\partial_y f_0(y_n)$ has large positive and negative eigenvalues, so that an application of $\left(I - \theta\tau\partial_y f_0(y_n)\right)^{-1}$ would imply a step size restriction as for explicit time integrators. Moreover, $\partial_y f_0(y_n)$ is not a banded matrix with small band-width. Then, the idea is to approximate the AMF factor

$$\left(I - \theta\tau\partial_y f_0(y_n)\right)^{-1} \approx I + \theta\tau\partial_y f_0(y_n) \prod_{j=1}^{m} \left(I - \theta\tau\partial_y f_j(y_n)\right)^{-1}. \tag{21}$$

We then have $\left(I - \theta\tau\partial_y f_0(y_n)\right)^{-1} \approx I + \theta\tau\partial_y f_0(y_n)$, but before applying the operator $\partial_y f_0(y_n)$ the large eigenvalues are damped by applying successively $\left(I - \theta\tau\partial_y f_j(y_n)\right)^{-1}$, $1 \le j \le m$. Hence, with the notation (17), PDE-W-methods are obtained with the following choice in (15)

$$(I - \theta \tau W)^{-1} = P_m^{-1} \left(I + \theta \tau \begin{pmatrix} 0 & 0 \\ a_{n,0} & A_{n,0} \end{pmatrix} P_m^{-1} \right),$$

$$P_m := \prod_{j=1}^m \begin{pmatrix} 1 & 0 \\ -\theta \tau a_{n,j} & (I - \theta \tau A_{n,j}) \end{pmatrix}.$$

(22)

The stages of the PDE-W method (A, L, b, θ) are computed for $i = 1, \ldots, s$, as follows:

$$K_i^{(0)} = \tau F(t_n + c_i \tau, U_n + \sum_{j=1}^{i-1} a_{ij} K_j) + \sum_{j=1}^{i-1} \ell_{ij} K_j$$

$$(I - \theta \tau A_{n,j}) K_i^{(j)} = K_i^{(j-1)} + \theta \rho_i \tau^2 a_{n,j}, \qquad (j = 1, \ldots, m)$$

$$\hat{K}_i^{(0)} = K_i^{(0)} + \theta \tau A_{n,0} K_i^{(m)} + \theta \rho_i \tau^2 a_{n,0}$$

(23)

$$(I - \theta \tau A_{n,j}) \hat{K}_i^{(j)} = \hat{K}_i^{(j-1)} + \theta \rho_i \tau^2 a_{n,j}, \qquad (j = 1, \ldots, m)$$

$$K_i = \hat{K}_i^{(m)},$$

with advancing solution after one step given by (20).

2.3 AMFR-W Methods

The AMFR-W methods have been recently introduced in [9] and they are based on the iteration

$$(I - \mu \tau V)(x^{(p)} - x^{(p-1)}) = \tilde{d} - (I - \theta \tau J) x^{(p-1)}, \qquad p = 1, 2, 3, \ldots \quad (24)$$

to solve linear systems of the form,

$$(I - \theta \tau J) x = \tilde{d}, \qquad \theta > 0, \qquad \tau > 0. \quad (25)$$

The convergence of this iteration for an arbitrary initial approximation $x^{(0)}$ is determined by the spectral radius $(\mathfrak{p})$ of the matrix $\mathscr{P}$ below,

$$\mathscr{P} = \tau (I - \mu \tau V)^{-1} (\theta J - \mu V), \qquad \mathfrak{p}(\mathscr{P}) < 1.$$

In [6, Sections 3-5] the authors followed this approach to derive some W-methods and other kind of AMF-W methods (see Theorems 5 and 11 in [6]) based on one- or two-stage ROW methods with orders of consistency two and three, respectively. Only two iterations per stage and $x^{(0)} = 0$ are needed to recover the full order of convergence of the underlying ROW method in both cases. Here, we follow the same approach for any W-method applied to parabolic problems with mixed derivatives, but treating the Jacobian for the mixed derivatives in an explicit way, i.e., it is not included in the AMF-factorization. It is also important to remark that this iteration introduces a new parameter μ, that will allow to improve the stability properties of the method.

The second iteration (which is a refinement of the first approximation $x^{(1)}$) in (24) by taking as initial guess $x^{(0)} = 0$ gives

$$x^{(2)} = (I - \theta \tau W)^{-1} \tilde{d},$$

where

$$(I - \theta \tau W)^{-1} = (I - \mu \tau V)^{-1} \left(2I - (I - \theta \tau J)(I - \mu \tau V)^{-1} \right). \tag{26}$$

It should be observed that this approach implies

$$J - W = \mathcal{O}(\tau), \quad \tau \to 0, \quad \forall \mu, \ \forall V.$$

The AMFR-W method can be seen as a refined AMF-W method and it is obtained from (15) with the choice made in (26) and (27)

$$I - \mu \tau V = \prod_{j=1}^{m} \begin{pmatrix} 1 & 0 \\ -\mu \tau a_{n,j} & (I - \mu \tau A_{n,j}) \end{pmatrix},$$

$$J = \begin{pmatrix} 0 & 0 \\ \partial_t F(t_n, U_n) & \partial_U F(t_n, U_n) \end{pmatrix}. \tag{27}$$

For a given ROW (A, L, b, θ), the corresponding AMFR-W method (A, L, b, θ, μ) is then given by the following formulation. For $i = 1, 2, \ldots, s$, compute K_i from:

$$
\begin{aligned}
K_i^{(0)} &= \tau F(t_n + c_i \tau, U_n + \sum_{j=1}^{i-1} a_{ij} K_j) + \sum_{j=1}^{i-1} \ell_{ij} K_j \\
(I - \mu \tau A_{n,j}) K_i^{(j)} &= K_i^{(j-1)} + \mu \rho_i \tau^2 a_{n,j}, \quad (j = 1, \ldots, m) \\
\hat{K}_i^{(0)} &= 2K_i^{(0)} + \theta \rho_i \tau^2 \partial_t F(t_n, U_n) - (I - \theta \tau \partial_U F(t_n, U_n)) K_i^{(m)}, \\
(I - \mu \tau A_{n,j}) \hat{K}_i^{(j)} &= \hat{K}_i^{(j-1)} + \mu \rho_i \tau^2 a_{n,j}, \quad (j = 1, \ldots, m) \\
K_i &= \hat{K}_i^{(m)}.
\end{aligned}
\tag{28}
$$

The numerical solution after one step is then computed from (20).

Remark 1 For the same number of stages, the implementation of PDE-W- and AMFR-W-methods requires a similar computational cost, and this is about twice the cost associated to AMF-W-methods.

3 Stability

The study of stability for W-methods gets complicated due to the fact that the matrices W and $\partial_y f(y_n)$ do not need to commute. In this section a scalar test equation is proposed that is relevant for a large class of partial differential equations for which the dominant part is an elliptic operator with constant coefficients endowed either with periodic boundary conditions [19] or homogeneous Dirichlet boundary conditions [10]. An analysis of unconditional stability for the families of methods presented in Sect. 2 on linear problems with constant coefficients is provided below. This stability analysis comprises some results that can also be found in [9, 10]. Both AMF-W- and AMFR-W-methods will be seen to be unconditionally stable regardless of the spatial dimension m at the expense of possibly increasing the stability parameters of the particular method. This aspect is shared with other classical ADI methods, like the Craig–Sneyd, Hundsdorfer–Verwer and the modified Craig–Sneyd schemes [19] whose temporal order of convergence is at most two. For PDE-W-methods, unconditional stability is only possible whenever $m \leq 3$, as it happens for the second order Craig–Sneyd scheme.

A standard second-order central space discretization of (12) leads to the linear ordinary differential equation

$$\dot{U} = \mathcal{M}U, \qquad U(0) = U_0, \tag{29}$$

where $\mathcal{M}$ is given by (13). The difficulty of studying the stability of time integrators lies in the fact that the differentiation matrices $D_{x_i} = \text{TriDiag}(-1, 0, 1)/(2\Delta x_i)$ and $D_{x_i x_i} = \text{TriDiag}(1, -2, 1)/\Delta x_i^2$ do not commute.

Theorem 1 *If the coefficient matrix $\mathscr{A} = (\alpha_{ij})_{i,j=1}^m$ in (12) is positive definite, then the system (29) is asymptotically stable.*

Proof The main idea is to approximate $D_{x_i x_i}$ by $D_{x_i}^2$ and to study the resulting defect. The matrix $\mathcal{M}$ can be split as $\mathcal{M} = \mathcal{M}_0 + \sum_{i=1}^m \mathcal{M}_i$ with

$$\mathcal{M}_i = \alpha_{i,i}(I_{n_{x_m}} \otimes \ldots \otimes \left(D_{x_i x_i} - D_{x_i}^2\right) \otimes \ldots \otimes I_{n_{x_1}}), \qquad i = 1, \ldots, m,$$

$$\mathcal{M}_0 = \sum_{i,j=1}^m \alpha_{i,j}(I_{n_{x_m}} \otimes \ldots \otimes D_{x_j} \otimes \ldots \otimes D_{x_i} \otimes \ldots \otimes I_{n_{x_1}}).$$

First, we observe the relation

$$D_{x_i x_i} = D_{x_i}^2 - \frac{\Delta x_i^2}{4} D_{x_i x_i}^2 - \frac{1}{2\Delta x_i^2}\text{Diag}\,(1, 0, \ldots, 0, 1),$$

which implies that the logarithmic norm of the defect $D_{x_i x_i} - D_{x_i}^2$ is negative for $i = 1, \ldots, m$.

Secondly, if we let v_i be an eigenvector of D_{x_i} (recall that the eigenvectors form an orthogonal base of $\mathbb{C}^{n_{x_i}}$, $n_{x_i} = 1/\Delta x_i - 1$) with eigenvalue $i\lambda_i$, then $v_m \otimes \ldots \otimes v_1$ is an eigenvector of $\mathcal{M}_0$ corresponding to the eigenvalue

$$\sum_{i,j=1}^{m} \alpha_{ij}(-\lambda_i \lambda_j) = -(\lambda_1, \ldots, \lambda_m)\mathcal{A}(\lambda_1, \ldots, \lambda_m)^\mathsf{T} \le 0.$$

This proves the asymptotic stability of the system (29). $\qquad\square$

Motivated by Theorem 1 we replace $D_{x_i x_i}$ by $D_{x_i}^2$ in $\mathcal{M}$, so that the system (29) can be decoupled into scalar linear ODEs of the form

$$\dot{u} = -\left(\sum_{i,j=1}^{m} \alpha_{i,j} \lambda_i \lambda_j\right)u, \quad \lambda_i \in \mathbb{R} \ (i = 1, \ldots, m) \tag{30}$$

where $i\lambda_i$ represents an eigenvalue of D_{x_i} and $\mathcal{A} = (\alpha_{i,j})_{i,j=1}^{m}$ is a symmetric positive definite matrix. Let us now consider the change

$$\lambda_i \leftrightarrow \lambda_i \sqrt{\alpha_{i,i}}, \quad c_{i,j} = \alpha_{i,j}/\sqrt{\alpha_{i,i} \cdot \alpha_{j,j}}, \quad \mathcal{C} = (c_{i,j})_{i,j=1}^{m} > 0, \tag{31}$$

which reduces the scalar test problem (30) to

$$\dot{u} = -\left(\sum_{i=1}^{m} \lambda_i^2 + 2 \sum_{1 \le i < j \le m} c_{i,j} \lambda_i \lambda_j\right)u, \quad \lambda_i \in \mathbb{R} \ (i = 1, \ldots, m). \tag{32}$$

Observe that the diagonal elements of the matrix $\mathcal{C} > 0$ satisfy $c_{i,i} = 1$, and the off-diagonal elements are bounded as $|c_{i,j}| < \sqrt{c_{i,i} \cdot c_{j,j}} = 1$ for $1 \le i, j \le m$, $i \ne j$.

Now, applying an AMF-type W-method to (32) with the splitting

$$F_j(t, u) = -\lambda_j^2 u \ (j = 1, \ldots, m), \quad F_0(t, u) = \left(-2 \sum_{i<j} c_{i,j} \lambda_i \lambda_j\right)u, \tag{33}$$

yields a recursion $u_{n+1} = R(z, z_1, \ldots, z_m)u_n$, where $R(z, z_1, \ldots, z_m)$ is a rational function of the real variables

$$z = z_0 + \sum_{j=1}^{m} z_j, \quad z_0 = -2\tau \sum_{1 \le i < j \le m} c_{i,j} \lambda_i \lambda_j, \quad z_j = -\tau \lambda_j^2, \quad 1 \le j \le m, \quad \tau > 0. \tag{34}$$

This rational function is called the *linear stability function* of the method. For an AMF-type W-method based on the coefficients (A, L, b, θ), it is given by

$$R(z, z_1, \ldots, z_m) = 1 + zb^\top \left(\tilde{\Pi}_m(\theta) I - L - zA \right)^{-1} \mathbf{1} \tag{35}$$

where $\mathbf{1} := (1, \ldots, 1)^\top \in \mathbb{R}^s$ and for each AMF approximation (18), (22) and (26)–(27) in Sect. 2, it holds that

(A) $\quad \dfrac{1}{\tilde{\Pi}_m(\theta)} = \dfrac{1}{\Pi_m(\theta)},$ for AMF-W-methods,

(B) $\quad \dfrac{1}{\tilde{\Pi}_m(\theta)} = \dfrac{1}{\Pi_m(\theta)} \left(1 + \dfrac{\theta z_0}{\Pi_m(\theta)} \right),$ for PDE-W-methods,

(C) $\quad \dfrac{1}{\tilde{\Pi}_m(\theta)} = \dfrac{1}{\Pi_m^*(\mu, \theta)},$ for AMFR-W-methods,

where $\quad \Pi_m(\theta) := \prod_{j=1}^m (1 - \theta z_j), \qquad \dfrac{1}{\Pi_m^*(\mu, \theta)} := \dfrac{1}{\Pi_m(\mu)} \left(2 - \dfrac{1 - \theta z}{\Pi_m(\mu)} \right).$

$$\tag{36}$$

Of course, in case (C), the AMF factor $\tilde{\Pi}_m(\theta)$ depends on the additional parameter $\mu > 0$.

It should be also noticed from (34) that $z \le 0$, for all $\lambda_j \in \mathbb{R}$, $\tau > 0$, because of the positive definiteness of the matrix $\mathscr{C}$.

Definition 1 A time integrator which, when applied to the test equation (32), yields the recursion $u_{n+1} = R(z, z_1, \ldots, z_m) u_n$ with stability function (35), is called unconditionally stable for a given $m \ge 2$, if

$$|R(z, z_1 \ldots, z_m)| \le 1$$

for all $z, z_1 \ldots, z_m$ of (34) and each matrix $\mathscr{C} > 0$.

The unconditional stability properties of AMF-type W-methods rely on the linear stability of the underlying ROW method (A, L, b, θ), whose linear stability function is obtained from (35) by replacing the AMF factor $\tilde{\Pi}_m(\theta)$ by $1 - \theta z$. Hence, it is given by

$$R_\theta(z) = 1 + zb^T ((1 - \theta z)I - L - zA)^{-1} \mathbf{1}, \quad z \in \mathbb{C}. \tag{37}$$

The ROW-method (A, L, b, θ) is A_0-stable when its stability function (37) fulfils

$$|R_\theta(x)| \le 1, \quad \text{for all } x \le 0.$$

The range of values for $\theta \geq \theta_0$ providing A_0-stable methods is known for many Rosenbrock methods. The most simple one-stage W-method is given by, (see e.g. [17, p. 398])

$$\left(I - \theta \tau W\right)(y_{n+1} - y_n) = \tau f(y_n).$$ (38)

It is of classical order 1, and for $\theta = 1/2$ reaches order 2 if (10) holds. Its stability function is

$$R_\theta(z) = 1 + \frac{z}{1 - \theta z},$$ (39)

and the method is A_0-stable whenever $\theta \geq 1/2$.

For two-stage methods, there is a two-parameter family of W-methods of order ≥ 2 (see [17, p. 400]) with free parameters θ, b_2 and coefficients given by

$$b_1 = 2 - b_2, \qquad a_{21} = \frac{1}{2b_2}, \qquad \ell_{21} = -\frac{1}{b_2}.$$ (40)

It is not difficult to check that the stability function for the methods (40) only depends on the stability parameter θ and is given by

$$R_\theta(z) = 1 + \frac{2z}{1 - \theta z} + \frac{z(z - 2)}{2(1 - \theta z)^2}$$ (41)

and A_0-stability is obtained as long as $\theta \geq 1/4$.

On the other hand, a family of 3-stage W-methods of order ≥ 3 under the special assumption (11) was studied in [7, Theorem 1]. There it was shown that, under the assumption (11), there exist two three-parametric families of 3-stage W-methods of order three, with free parameters a_{32}, a_{21} and θ, satisfying $a_{32}a_{21} \neq 0$, whose coefficients are given by

$$b_3 = \frac{1}{6a_{32}a_{21}}, \qquad \ell_{21} = r a_{21}, \qquad \ell_{32} = \frac{6}{r}a_{32},$$
$$a_{31}^2 + (-a_{21} + 2a_{32} + 2r a_{32}a_{21}) a_{31}$$
$$+ \left[a_{32}(a_{32} + a_{21}(-3 + 2r a_{32}) + a_{21}^2(6/r + 9 + r^2 a_{32}))\right] = 0,$$
$$b_2 = \frac{3/2 - b_3(a_{31} + a_{32})}{a_{21}}, \qquad \ell_{31} = -\frac{3 + b_2\ell_{21} + b_3\ell_{32}}{b_3}, \qquad b_1 = 3 - (b_2 + b_3)$$ (42)

and $r = -3 \pm \sqrt{3}$. All the methods in (42) have the same stability function (depending only on θ), which is given by

$$R_\theta(z) = 1 + \frac{3z}{1 - \theta z} + \frac{3z(z - 2)}{2(1 - \theta z)^2} + \frac{z(z^2 - 6z + 6)}{6(1 - \theta z)^3}.$$ (43)

In this case, A_0-stability is obtained for $\theta \geq 1/3$. It must also be observed that there do not exist 3-stage W-methods of order 3 without any restriction on W [27].

Further, a one-parameter family of four-stage fourth-order ROW-methods with the classical Kutta's 3/8-rule method as underlying explicit Runge-Kutta method was introduced in [8, Section 6]. Its coefficients (9) are given by $A = \widetilde{A}\Gamma^{-1}$, $L = I_4 - \Gamma^{-1}$ and $b^T = \widetilde{b}^T \Gamma^{-1}$, with

$$
\Gamma = \begin{pmatrix} 1 & 0 & 0 & 0 \\ -\frac{4}{3} & 1 & 0 & 0 \\ -\frac{2(-1-2\theta+12\theta^2)}{3(-1+4\theta)(-1+6\theta)} & -\frac{2(1-6\theta+12\theta^2)}{(-1+4\theta)(-1+6\theta)} & 1 & 0 \\ \frac{24\theta(-1+3\theta)}{(-1+4\theta)(-1+6\theta)} & \frac{6(1-6\theta+12\theta^2)}{(-1+4\theta)(-1+6\theta)} & -6 & 1 \end{pmatrix},
\tag{44}
$$

$$
\widetilde{A} = \begin{pmatrix} 0 & 0 & 0 & 0 \\ \frac{1}{3} & 0 & 0 & 0 \\ -\frac{1}{3} & 1 & 0 & 0 \\ 1 & -1 & 1 & 0 \end{pmatrix}, \qquad \widetilde{b}^T = \left(\frac{1}{8}, \frac{3}{8}, \frac{3}{8}, \frac{1}{8} \right)^T.
$$

The associated linear stability function is then given by

$$
R_\theta(z) = 1 + \frac{q_1 z}{1-\theta z} + \frac{q_2 z}{(1-\theta z)^2} + \frac{q_3 z}{(1-\theta z)^3} + \frac{q_4 z}{(1-\theta z)^4}, \qquad \text{with}
$$

$$
q_1 = \frac{96\theta^2 - 42\theta + 5}{(4\theta-1)(6\theta-1)}, \qquad q_3 = \frac{(21-138\theta+288\theta^2-14z+120\theta z-276\theta^2 z+2z^2-19\theta z^2+45\theta^2 z^2)}{(3(4\theta-1)(6\theta-1))},
$$

$$
q_2 = \frac{210\theta^2 z - 432\theta^2 - 90\theta z + 198\theta + 10z - 27}{3(4\theta-1)(6\theta-1)}, \quad q_4 = \frac{(z-6)(z-4)(24\theta^2 z - 24\theta^2 - 10\theta z + 12\theta + z - 2)}{24(4\theta-1)(6\theta-1)},
$$

$$
\tag{45}
$$

and A_0-stability is obtained whenever $\theta \geq (3+\sqrt{3})/12$.

The A_0-stability properties of the methods (38), (40), (42) and (44) above are collected in Table 2, where s denotes the number of stages, and they have order of consistency $p \geq s$.

Observe that the linear stability functions (35) for the different AMF options (36) are obtained from (39)–(45) by replacing the factor $1 - \theta z$ for the corresponding AMF factor $\widetilde{\Pi}_m(\theta)$.

Table 2 Values of $\theta \geq \theta_0$ for some s-stage A_0-stable ROW methods (A, L, b, θ) of order $p \geq s$.

Method	$s = 1$	$s = 2$	$s = 3$ (42)	$s = 4$ (44)
θ_0	1/2	1/4	1/3	$(3+\sqrt{3})/12$

The following assumption (46) is the main ingredient to prove unconditional stability for AMF-type W-methods. It relates the stability of such a method with the A_0-stability of the underlying ROW method. Here, θ^* is some constant which may depend on θ and m.

$$0 < \frac{1}{\tilde{\Pi}_m(\theta)} \le \frac{1}{1 - \theta^* z}, \quad \theta^* \ge 0, \quad \forall \mathscr{C} > 0, \quad \forall z, z_1 \ldots, z_m \text{ in (34)}. \tag{46}$$

Theorem 2 *Assume that the ROW method (A, L, b, θ) is A_0-stable for any $\theta \ge \theta_0 > 0$, and consider an AMF-type W-method with stability function given by (35)–(36). If for the given θ (and μ in case (C)) (46) holds with $\theta^* \ge \theta_0$, then the AMF-type W-method is unconditionally stable.*

Proof Since $z \le 0$, taking into account (46) and the mean value theorem, it holds that

$$\frac{1}{\tilde{\Pi}_m(\theta)} = \frac{1}{1 - \nu z} \tag{47}$$

for some $\nu \ge \theta^*$ that may depend on θ, z, and z_j, $j = 1, \ldots, m$. Therefore,

$$R(z, z_1 \ldots, z_m) = R_\nu(z).$$

Now, the proof is concluded from the A_0−stability of the ROW method together with $\nu \ge \theta^* \ge \theta_0$. $\qquad\square$

Theorem 2 will be considered in forthcoming subsections in order to show unconditional stability for the AMF-type W-methods presented in Sect. 2. The application of this result will depend on whether the assumption (46) is fulfilled for such methods.

3.1 Stability of AMF-W-Methods

Theorem 3 *For the stability function of the AMF-W-method (35) and (36)-(A) we have that (46) holds with $\theta^* = m^{-1}\theta > 0$.*

Proof The left inequality in (46) is trivial. To show the one on the right, let us define the vector

$$v := (|y_1|, |y_2|, \ldots, |y_m|)^T, \quad \text{with} \quad y_j = \sqrt{\theta \tau} \lambda_j, \quad 1 \le j \le m.$$

Since $|c_{i,j}| \leq 1$, $1 \leq i, j \leq m$, it then holds that

$$\tilde{\Pi}_m(\theta) - (1 - m^{-1}\theta z) = \prod_{j=1}^{m}(1 + y_j^2) - 1 - m^{-1}\sum_{i,j=1}^{m} c_{i,j} y_i y_j$$
$$\geq (1 + \sum_{j=1}^{m} y_j^2) - 1 - m^{-1}\sum_{i,j=1}^{m} |y_i||y_j|$$
$$= \sum_{j=1}^{m} y_j^2 - m^{-1}(\sum_{j=1}^{m} |y_j|)^2$$
$$\geq \|v\|_2^2 - m^{-1}m\|v\|_2^2 = 0,$$

where the last inequality above follows from the Cauchy-Schwarz inequality. □

Observe that the value $\theta^* = m^{-1}\theta$ given in Theorem 3 is optimal since a right inequality as in (46) for other θ^* such that $0 \leq \theta^* < m^{-1}\theta$ cannot be obtained.

Corollary 1 *Assume that the ROW method (A, L, b, θ) is A_0-stable for any $\theta \geq \theta_0 > 0$. Then, the AMF-W method (A, L, b, θ) (19)–(20) is unconditionally stable as long as $\theta \geq m\theta_0$.*

Proof It follows directly from Theorems 2 and 3. □

The previous result allows to show the unconditional stability of several important AMF-W methods considered in the literature.

Theorem 4 *Consider a family of s-stage consistent AMF-W methods (A, L, b, θ) (19)–(20).*

1. *For $s = 1$, the methods are unconditionally stable as long as $\theta \geq m/2$.*
2. *For $s = 2$ and order of consistency at least two for the underlying ROW methods, the corresponding AMF-W methods are unconditionally stable as long as $\theta \geq m/4$.*
3. *For $s = 3$, the AMF-W methods with coefficients given in (42) are unconditionally stable as long as $\theta \geq m/3$.*
4. *For $s = 4$, the AMF-W methods based on the Kuttas's 3/8-rule with coefficients given in (44) are unconditionally stable as long as $\theta \geq m\theta_0$, with $\theta_0 = (3 + \sqrt{3})/12$.*

Proof The θ-values that provide A_0-stability for the family of s-stage consistent ROW methods (A, L, b, θ), with $\theta \geq \theta_0$, are given in Table 2. The proof now follows from Corollary 1. □

3.2 Stability of PDE-W-Methods

The properties of unconditional stability of PDE-W-methods were studied in [10], where it was seen that there exist unconditionally stable methods for $m = 2, 3$ but not for $m \geq 4$ on arbitrary linear parabolic problems with constant coefficients (12).

Anyhow, let us first show that the right inequality in (46) holds with $\theta^* = \theta$ for all $m \geq 2$.

Theorem 5 *Let $\lambda_i \in \mathbb{R}$, $1 \leq i \leq m$, and assume that $\mathscr{C} = (c_{i,j})_{i,j=1}^m$ (with $c_{i,i} = 1$) is positive definite. With z_i of (34) and $z = z_0 + z_1 + \ldots + z_m$, then the AMF factor $\tilde{\Pi}_m(\theta)$ (with $\theta \geq 0$) given by (36)-(B) satisfies*

$$\frac{1}{\tilde{\Pi}_m(\theta)} \leq \frac{1}{1 - \theta z}. \tag{48}$$

Proof Let us take $w_i = -\theta z_i$ for $0 \leq i \leq m$, in such a way that $w_0 \in \mathbb{R}$, $w_i \geq 0$, $1 \leq i \leq m$, satisfy $1 + w_0 + \sum_{i=1}^m w_i > 0$. Then (48) is equivalent to show that

$$\frac{1}{\prod_{i=1}^m (1 + w_i)} \left(1 - \frac{w_0}{\prod_{i=1}^m (1 + w_i)}\right) \leq \frac{1}{1 + w_0 + \sum_{i=1}^m w_i}.$$

Let us define $P_m := \prod_{i=1}^m (1 + w_i) \geq 1$ and $S_m := \sum_{i=1}^m w_i \geq 0$. Then, it is not difficult to check that

$$\frac{1}{P_m}\left(1 - \frac{w_0}{P_m}\right) - \frac{1}{1 + w_0 + S_m} = \frac{-\left(w_0 + \frac{1}{2}(1 + S_m - P_m)\right)^2 + \frac{1}{4}(1 + S_m + P_m)^2 - P_m^2}{P_m^2(1 + w_0 + S_m)}.$$

In order to show that this expression is non positive, observe that

$$\frac{1}{4}(1 + S_m + P_m)^2 - P_m^2 = \frac{1}{4}\left(1 + S_m + 3P_m\right)\left(1 + S_m - P_m\right) \leq 0$$

since $P_m \geq 1 + S_m$. This concludes the proof. □

The stability analysis of PDE-W-methods also requires the positivity of $\tilde{\Pi}_m(\theta)$ so that the left inequality in (46) is fulfilled. However, this condition can only be satisfied for all positive definite matrices $\mathscr{C}$ as long as $m = 2, 3$. To see this, we rewrite the condition $\tilde{\Pi}_m(\theta) > 0$ as

$$\prod_{j=1}^m (1 + y_j^2) - \sum_{i \neq j} c_{i,j} y_i y_j > 0 \qquad \text{for all } y_i \in \mathbb{R}. \tag{49}$$

Considering the change $y_i = \sqrt{\theta \tau} \lambda_i$, this inequality becomes equivalent to the positivity of the factor $\tilde{\Pi}_m(\theta)$ of (36)-(B). Unconditionl stability for PDE-W-methods then requires (49) to hold for all positive definite matrices $\mathscr{C}$. However, it turns out that this is true in dimensions $m = 2, 3$, but not in general for $m \geq 4$. Observe that for $m = 2$, (49) follows immediately since $|c_{1,2}| < 1$. For $m = 3$, we have the following result.

Theorem 6 *Assume that $\mathscr{C} = (c_{i,j})_{i,j=1}^{3}$ is positive definite, with $c_{i,i} = 1$ for all i. Then, (49) holds with $m = 3$.*

Proof First observe that $|c_{i,j}| < \sqrt{c_{i,i} \cdot c_{j,j}} = 1$ $(1 \le i, j \le 3, \ i \ne j)$. Then, for all $y_j \in \mathbb{R}$

$$\prod_{j=1}^{3}(1 + y_j^2) - \sum_{i \ne j} c_{i,j} y_i y_j \ge 1 + \sum_{j=1}^{3} y_j^2 + \sum_{i<j} y_i^2 y_j^2 - \sum_{i \ne j} |y_i||y_j|$$

$$\ge -2 + \sum_{j=1}^{3} y_j^2 + \sum_{i<j} (|y_i||y_j| - 1)^2.$$

Let us now consider the function

$$f(y_1, y_2, y_3) = -2 + (y_1^2 + y_2^2 + y_3^2) + (y_1 y_2 - 1)^2 + (y_1 y_3 - 1)^2 + (y_2 y_3 - 1)^2,$$

with $y_1, y_2, y_3 \ge 0$. It is then seen that the critical points of $f(y_1, y_2, y_3)$ fulfil $y_1 = y_2 = y_3$, and the minimum value of f is $f(\frac{1}{\sqrt{2}}, \frac{1}{\sqrt{2}}, \frac{1}{\sqrt{2}}) = \frac{1}{4} > 0$. $\qquad\square$

Remark 2 The previous result for dimension $m = 3$ is not true in higher dimensions, i.e., (49) does not hold for arbitrary positive definite matrices $\mathscr{C} = (c_{i,j})_{i,j=1}^{m}$, with $c_{i,i} = 1$ $(1 \le i \le m)$ whenever $m \ge 4$. This can be seen by taking $y_j = y \ge 0$ $(1 \le j \le m)$ in (49) and considering

$$f(y) = (1 + y^2)^m - Sy^2, \qquad S = \sum_{i \ne j} c_{i,j}.$$

If $S \ge m$, $f(y)$ attains a minimum at the point $y^* \ge 0$, with

$$(y^*)^2 = -1 + \left(\frac{S}{m}\right)^{1/(m-1)}.$$

For this value, one has $f(y^*) > 0$ if and only if

$$S = \sum_{i \ne j} c_{i,j} < m \left(\frac{m}{m-1}\right)^{m-1}. \tag{50}$$

Hence (50) is a necessary condition for (49).

Remark 3 A sufficient condition for (49) to hold in dimension $m \ge 4$ is that the matrix $2I - \mathscr{C}$ is positive semi-definite. In fact, expanding the product in (49) and neglecting the fourth and higher order terms shows that (49) holds if

$$y_1^2 + \ldots + y_m^2 - \sum_{i \ne j} c_{i,j} y_i y_j \ge 0, \tag{51}$$

and, since $c_{i,i} = 1$, this is equivalent to $2I - \mathscr{C} \ge 0$.

The stability of PDE-W-methods in dimension $m \geq 2$ can be now established under the assumption (49).

Corollary 2 *Assume that the ROW method (A, L, b, θ) is A_0-stable for any $\theta \geq \theta_0 > 0$. If (49) holds, then the PDE-W method (A, L, b, θ) (23)–(20) is unconditionally stable as long as $\theta \geq \theta_0$.*

Proof Equation (46) follows from Theorem 5 and assumption (49). Now the proof is a consequence of Theorem 2 with $\theta^* = \theta$. □

Theorem 7 *Consider a family of s-stage consistent PDE-W methods (A, L, b, θ) (23)–(20). Under the assumption (49),*

1. *for $s = 1$, the methods are unconditionally stable as long as $\theta \geq 1/2$.*
2. *For $s = 2$ and order of consistency at least two for the underlying ROW methods, the corresponding PDE-W methods are unconditionally stable as long as $\theta \geq 1/4$.*
3. *For $s = 3$, the PDE-W methods with coefficients given in (42) are unconditionally stable as long as $\theta \geq 1/3$.*
4. *For $s = 4$, the PDE-W methods based on the Kuttas's 3/8-rule with coefficients given in (44) are unconditionally stable as long as $\theta \geq (3 + \sqrt{3})/12$.*

Proof The proof now follows from Corollary 2 and the θ-values in Table 2 that provide A_0-stability for the family of s-stage consistent ROW methods (A, L, b, θ), with $\theta \geq \theta_0$. □

3.3 Stability of AMFR-W-Methods

For each integer $m \geq 2$ let us consider the polynomial

$$g_m(x) = 2x \left(\frac{m - x}{m - 1} \right)^{m-1} - 1, \tag{52}$$

and denote by κ_m the smallest positive zero of $g_m(x)$. These numbers play a relevant role in the stability analysis for AMFR-W-methods as Theorems 8 and 9 below reflect. We first state a Lemma showing some properties of these zeros.

Lemma 1

1. *Let $g(x) := 2x \exp(1 - x) - 1$, with $x \geq 0$. Then*

$$g(x) > 0 \iff x \in (\kappa^*, \mathscr{K}^*), \quad \text{with } \kappa^* = 0.2319\ldots, \quad \mathscr{K}^* = 2.6783\ldots \tag{53}$$

2. *Let $g_m(x) := 2x \left(\frac{m-x}{m-1} \right)^{m-1} - 1$, with $x \in [0, m]$, $m \in \mathbb{N}$, $m \geq 2$. Then*

$$g_m(x) > 0 \iff x \in (\kappa_m, \mathscr{K}_m), \quad \text{with } \kappa^* < \kappa_m < \mathscr{K}_m < \mathscr{K}^*. \tag{54}$$

3. *For all $m \in \mathbb{N}$, $m \geq 2$, it holds that $\kappa_{m+1} < \kappa_m$ and $\mathcal{K}_m < \mathcal{K}_{m+1}$.*
4. *For all $m \in \mathbb{N}$, $m \geq 2$, it holds that*

$$(m+1)\kappa_{m+1} > m\kappa_m. \tag{55}$$

Proof First, in order to prove items 1–3, we observe that the functions $g_m(x) = 2x\left(\frac{m-x}{m-1}\right)^{m-1} - 1$, $x \in [0, m]$, $m \geq 2$, fulfil $g_m(0) = g_m(m) = -1$, $g_m(1) = 1$ and $g'_m(x) > 0$, for $x \in (0, 1)$, and $g'_m(x) < 0$, for $x \in (1, m)$. This shows that there exist real numbers $0 < \kappa_m < 1 < \mathcal{K}_m < m$ such that

$$g_m(x) > 0 \iff x \in (\kappa_m, \mathcal{K}_m).$$

Moreover, it is readily checked that $g_m(1/2) > 0$, for all $m \geq 2$. Hence, $\kappa_m < \frac{1}{2}$, for all $m \geq 2$.

Now, we observe that $g_{m+1}(x) > g_m(x)$, $\forall x \in (0, m)$, $x \neq 1$, and from here it holds that $g_{m+1}(\kappa_m) > 0$ and $g_{m+1}(\mathcal{K}_m) > 0$, which implies $\kappa_{m+1} < \kappa_m$ and $\mathcal{K}_m < \mathcal{K}_{m+1}$. To see that $g_{m+1}(x) > g_m(x)$, $\forall x \in (0, m)$, $x \neq 1$, it is enough to consider that

$$g_{m+1}(x) - g_m(x) = (2x)\left(\left(1 + \frac{1-x}{m}\right)^m - \left(1 + \frac{1-x}{m-1}\right)^{m-1}\right) > 0$$

since

$$\frac{\left(1 + \frac{1-x}{m}\right)^m}{\left(1 + \frac{1-x}{m-1}\right)^{m-1}} = \left(1 - \frac{1-x}{m(m-x)}\right)^m \left(\frac{m-x}{m-1}\right)$$

$$> \left(1 - (m)\frac{1-x}{m(m-x)}\right)\left(\frac{m-x}{m-1}\right) = 1$$

for all $\forall x \in (0, m)$, $x \neq 1$, by virtue of the Bernoulli's inequality.

The proof of items 1-3 is concluded taking into account that the function $g(x) = (2x)\exp(1 - x) - 1$ is the pointwise limit of $g_m(x)$ as $m \to \infty$, for all $x \geq 0$.

To prove item 4, we shall next check that $g_m\left(\frac{m+1}{m}x\right) > g_{m+1}(x)$, for all $m \geq 2$ and $x \in (0, \frac{1}{2})$. Hence, the proof of item 4 follows just by evaluating this inequality at $x = \kappa_{m+1}$ and considering the property stated in item 2.

To see that $g_m\left(\frac{m+1}{m}x\right) > g_{m+1}(x)$, for all $m \geq 2$ and $x \in (0, \frac{1}{2})$, from a direct calculation we have that

$$g_m\left(\frac{m+1}{m}x\right) - g_{m+1}(x) = \frac{2x}{(m(m-1))^m}h_m(x),$$

with $h_m(x) = (m^2 - 1)(m^2 - (m+1)x)^{m-1} - ((m^2 - 1) - (m-1)x)^m$. Now we have for $m \geq 2$ and $x \in (0, \frac{1}{2})$ that

$$h_m(x) > ((m^2 - 1) - (m-1)x) \left\{ (m^2 - (m+1)x)^{m-1} - ((m^2 - 1) - (m-1)x)^{m-1} \right\}$$

and the expression on the right hand side is positive since for all $x \in (0, \frac{1}{2})$ we have that $(m^2 - (m+1)x) - ((m^2 - 1) - (m-1)x) = 1 - 2x > 0$. □

Theorem 8 Let $m \geq 2$ be any given integer and $\theta > 0$. If $\mu \geq m\kappa_m\theta$ then for the stability function of the AMFR-W method (35) and (36)-(C) we have that (46) holds with $\theta^* = \theta$, i.e., for $\Pi_m^*(\mu, \theta)$ defined in (36)-(C), it holds that

$$0 < \frac{1}{\Pi_m^*(\mu, \theta)} \leq \frac{1}{1 - \theta z}, \qquad \forall \mathscr{C} > 0, \qquad \forall z, z_1 \ldots, z_m \text{ in (34)}.$$

Proof The inequality on the right follows immediately taking into account that

$$\frac{1}{1 - \theta z} - \frac{1}{\Pi_m^*(\mu, \theta)} = \frac{1}{1 - \theta z} \left(1 - \frac{1 - \theta z}{\Pi_m(\mu)} \right)^2 \geq 0.$$

The positivity of $\Pi_m^*(\mu, \theta)$ is equivalent to show that $2\Pi_m(\mu) - (1 - \theta z) > 0$, and considering the change of variables $y_j = \sqrt{\mu\tau}\lambda_j$, this can be written as

$$\mathscr{D} := 2 \prod_{j=1}^m \left(1 + y_j^2 \right) - 1 - \frac{\theta}{\mu} \sum_{i,j=1}^m c_{i,j}\, y_i y_j > 0.$$

Using $|c_{i,j}| \leq 1$, we obtain the lower bound

$$\mathscr{D} \geq 2 \prod_{j=1}^m \left(1 + y_j^2 \right) - 1 - \frac{\theta}{\mu} \left(\sum_{j=1}^m |y_j| \right)^2. \tag{56}$$

It follows from Lemma 2 below that this lower bound is non-negative, if $\mu \geq m\kappa_m\theta$, and that it can be equal to 0 only if $y_1 = \ldots = y_m = y$ for some $y \neq 0$. However, in this latter case the inequality in (56) is strict. This completes the proof. □

Lemma 2 Let $m \geq 2$ be any given integer and κ_m given by (54). If $\delta \leq \dfrac{1}{m\kappa_m}$ then it holds that

$$2 \prod_{j=1}^m (1 + y_j^2) - 1 - \delta \left(\sum_{j=1}^m y_j \right)^2 \geq 0, \qquad \forall\, y_j \geq 0, \ 1 \leq j \leq m. \tag{57}$$

Moreover, the equality can only hold if and only if $y_1 = y_2 = \ldots = y_m = y$, for some $y \neq 0$.

Proof First observe from (53)-(54) that $m\kappa_m > \frac{1}{2}$, for all $m \geq 2$, and hence $\delta < 2$. Let us now define, for $y_j > 0$, $1 \leq j \leq m$,

$$f(y_1, \ldots, y_m) := 2\prod_m - 1 - \delta\left(\sum_m\right)^2, \quad \text{with} \quad \prod_m := \prod_{j=1}^m (1 + y_j^2), \quad \sum_m = \sum_{j=1}^m y_j.$$

Since $\dfrac{\partial f}{\partial y_i} = \dfrac{4y_i}{1 + y_i^2} \prod_m - 2\delta \sum_m$, it follows that the components of a critical point $\mathbf{y} = (y_1, \ldots, y_m)$ for f must fulfil

$$\frac{y_i}{1 + y_i^2} = \frac{\delta \sum_m}{2 \prod_m}, \quad 1 \leq i \leq m.$$

Since $\dfrac{a}{1 + a^2} = \dfrac{b}{1 + b^2}$ implies $a = b$ or $ab = 1$, a critical point $\mathbf{y} = (y_1, \ldots, y_m)$ for f must fulfil that

$$\forall i, j \in \{1, \ldots, m\}, \ i \neq j : y_i = y_j \quad \text{or} \quad y_i = \frac{1}{y_j}.$$

We shall now see that all components y_j must be equal. To this aim, let us assume that a critical point $\mathbf{y}$ has m_1 components equal to $y > 1$ and $m - m_1$ components equal to $\dfrac{1}{y}$, for a certain m_1, $1 \leq m_1 \leq m - 1$. Then $\prod_m = (1 + y^2)(1 + \frac{1}{y^2})\hat{\prod}_m$ and $\sum_m = y + \frac{1}{y} + \hat{\sum}_m$, with

$$\hat{\prod}_m = \prod_{k=1}^{m_1-1}(1 + y^2) \prod_{k=1}^{m-m_1-1}(1 + \frac{1}{y^2}) \geq 1 + (m_1 - 1)y^2 + (m - m_1 - 1)\frac{1}{y^2}$$

and

$$\hat{\sum}_m = \sum_{k=1}^{m_1-1} y + \sum_{k=1}^{m-m_1-1} \frac{1}{y} = (m_1 - 1)y + (m - m_1 - 1)\frac{1}{y}.$$

Since $\delta < 2$, we would have

$$\begin{aligned}
\frac{\partial f}{\partial y_i}(\mathbf{y}) &= \frac{4y}{1 + y^2} \prod_m - 2\delta \sum_m \\
&> 4\left(\frac{y}{1 + y^2} \prod_m - \sum_m\right) \\
&\geq 4\left((m_1 - 1)y^3 + (m - m_1 - 1)\frac{1}{y^3}\right) \geq 0,
\end{aligned}$$

and hence $\dfrac{\partial f}{\partial y_i}(\mathbf{y}) > 0$, which contradicts that $\mathbf{y}$ is a critical point for f. Therefore, we must have $m_1 = 0$ or $m_1 = m$, that is, a critical point for f must have equal components.

Let then $\mathbf{y} = (y, \ldots, y) \in \mathbb{R}^m$ be such a critical point. It must then hold that

$$4y(1 + y^2)^{m-1} - 2m\delta y = 0$$

and from here it is readily seen that for $\delta \le \dfrac{2}{m}$ the only critical point is $(0, \ldots, 0)$, for which $f(0, \ldots, 0) = 1$. On the other hand, for $\delta > \dfrac{2}{m}$ there is also a nontrivial critical point $\mathbf{y} = (y, \ldots, y) \in \mathbb{R}^m$ with

$$(1 + y^2)^{m-1} = \frac{m\delta}{2}.$$

By making the change $\delta \leftrightarrow \kappa$, $\delta := \dfrac{1}{m \cdot \kappa}$, it is readily seen that $\dfrac{2}{m} < \delta \le \dfrac{1}{m\kappa_m}$ gives $\kappa_m \le \kappa < \dfrac{1}{2}(< \mathcal{K}_m)$. Hence, $\kappa \in [\kappa_m, \mathcal{K}_m)$.

Furthermore, using the fact that $(1 + y^2)^m = \left(\dfrac{1}{2\kappa}\right)^{\frac{m}{m-1}}$, a simple calculation shows that

$$f(y, \ldots, y) = 2(1 + y^2)^m - 1 - \frac{m}{\kappa}y^2$$

$$= 2(1 + y^2)^m - 1 - \frac{m}{\kappa}\frac{(1 + y^2)^m}{\frac{m\delta}{2}} + \frac{m}{\kappa}$$

$$= \frac{1}{\kappa}\left(2\kappa(1 + y^2)^m(1 - m) + (m - \kappa)\right)$$

$$= \frac{1}{(m - 1)\kappa(2\kappa)^{\frac{1}{m-1}}}\left(-1 + (2\kappa)^{\frac{1}{m-1}}\left(\frac{m - \kappa}{m - 1}\right)\right) \ge 0,$$

by virtue of Lemma 1 since $\kappa \in [\kappa_m, \mathcal{K}_m)$.

Finally, an induction argument shows that $f(y_1, y_2, \ldots, y_m) > 0$ in case that $y_j = 0$ for some $j \in \{1, 2, \ldots, m\}$. In this case, it must be observed that from (55) it holds that

$$\delta \le \frac{1}{m\kappa_m} < \frac{1}{(m - 1)\kappa_{m-1}} < \ldots < \frac{1}{2\kappa_2}.$$

□

Remark 4 A natural option to select the additional parameter μ in AMFR-W-methods is $\mu = 0$. In that case unconditional stability holds in dimensions $m = 2$

Table 3 Values for κ_m and $\mathcal{K}_m$ in Lemma 1, $2 \leq m \leq 9$, rounded up with 4 digits

m	2	3	4	5	6	7	8	9
κ_m	0.2929	0.2680	0.2576	0.2519	0.2482	0.2457	0.2439	0.2425
$\mathcal{K}_m$	1.7071	2	2.1572	2.2552	2.3223	2.3709	2.4079	2.4370

and $m = 3$, since $m\kappa_m < 1$ for $m \leq 3$. However, for $m \geq 4$, in order to guarantee unconditional stability one has to take $\mu \geq m\kappa_m\theta > \theta$.

Corollary 3 *Assume that the ROW method (A, L, b, θ) is A_0-stable for any $\theta \geq \theta_0 > 0$. Then, the AMFR-W method (A, L, b, θ, μ) (28)-(20) is unconditionally stable as long as $\mu \geq \kappa_m m\theta$, with $\theta \geq \theta_0$.*

Proof From Theorem 8 we get $\theta^* = \theta$ in (46) as long as $\mu \geq \kappa_m m\theta$. The result then follows from Theorem 2.   □

Theorem 9 *Consider a family of s-stage consistent ROW methods (A, L, b, θ). If $\mu \geq \kappa_m \, m\theta$ and $\theta \geq \theta_0$, then all consistent AMFR-W methods (A, L, b, θ, μ) are unconditionally stable*

1. *for $s = 1$ and $\theta_0 = 1/2$.*
2. *For $s = 2$ with order of consistency at least two as ROW method and $\theta_0 = 1/4$.*
3. *For $s = 3$, $\theta_0 = 1/3$ and the family of ROW methods with coefficients given in (42).*
4. *For $s = 4$, $\theta_0 = (3 + \sqrt{3})/12$ and the family of ROW methods based on the Kuttas's 3/8-rule considered in (44).*

Proof It follows from Corollary 3, taking into account that the θ-values that provide A_0-stability for the family of consistent ROW-methods (A, L, b, θ), with $\theta \geq \theta_0$, depending on the number of stages s, are given in Table 2.   □

Remark 5 For all $m \geq 2$, the stability bounds in Table 1 for the Hundsdorfer–Verwer scheme (4) coincide with the corresponding ones given in Theorem 9 for the one-stage AMFR-W method with $\theta = 1/2$, that is, $\mu_m = m\kappa_m\theta$ with κ_m in Table 3.

4 Numerical Experiments

The AMF-type W-methods of Sect. 2 will be compared to classical ADI schemes like the Hundsdorfer–Verwer and the modified Craig–Sneyd schemes (4)–(5) in the time integration of a linear diffusion model with constant coefficients in three and four spatial dimensions ($m = 3, 4$) and the 2D Heston model ($m = 2$) from finance. Fixed stepsize integrations are considered in Figs. 1, 2, 3, 4, 5, and 6 below so as to check unconditional stability and observe the temporal order of convergence in the ℓ^2-norm. The efficiency of the time integrators presented below is measured in relation to CPU time versus global errors. Moreover, each figure contains dashed straight lines with slopes two and three, respectively, to compare the temporal orders of convergence for the methods under consideration. Additional numerical experiments on the above-mentioned problems can also be found in [9].

HV is the method (4) with parameters $\mu = 1/2$ and $\theta > 0$ to be selected for stability requirements. The method is order two in ODE sense and it is unconditionally stable for $\theta \geq 1 - \frac{\sqrt{2}}{2}$, $\theta \geq 0.4020$ and $\theta \geq 0.5152$ when $m = 2$, $m = 3$ and $m = 4$, respectively (see Table 1). For $2 \leq m \leq 4$ we shall consider $\theta = (3 + \sqrt{3})/6$, $\theta = 0.4020$ and $\theta = 0.5152$, respectively.

MCS is the method (5) with parameters $\sigma = \theta$, $\mu = 1/2 - \theta$ and $\theta > 0$ to be chosen for stability. This scheme is order two in ODE sense and it is unconditionally stable for $\theta \geq \frac{1}{3}$, $\theta \geq \frac{6}{13}$ and $\theta \geq \frac{54}{91}$ when $m = 2$, $m = 3$ and $m = 4$, respectively (see Table 1). For $2 \leq m \leq 4$ we shall consider $\theta = \frac{1}{3}$, $\theta = \frac{6}{13}$ and $\theta = \frac{54}{91}$, respectively.

AMFR-W1 is the 1-stage AMFR-W-method (A, L, b, θ, μ) with coefficients

$$A = L = 0, \quad b = 1. \tag{58}$$

where we have chosen $\mu = \theta = 1/2$ for $m \leq 3$ and $\mu = 4\kappa_4\theta$, $\theta = 1/2$ for $m = 4$ (with $\kappa_4 = 0.2576$) to meet the stability bounds given in Theorem 9. This method is order two in ODE sense.

AMF-W2 is the 2-stage AMF-W-method (A, L, b, θ) with coefficients taken from [17, p. 155]

$$A = \begin{pmatrix} 0 & 0 \\ 2/3 & 0 \end{pmatrix}, \quad L = \begin{pmatrix} 0 & 0 \\ -4/3 & 0 \end{pmatrix}, \quad b = \begin{pmatrix} 5/4 \\ 3/4 \end{pmatrix}. \tag{59}$$

We have chosen $\theta = (3 + \sqrt{3})/6$ for $m \leq 3$ and $\theta = 1$ for $m = 4$ to ensure stability according the stability bounds given in Theorem 4. The method is only order two in ODE sense since (10) is not satisfied.

PDE-W2 is the 2-stage PDE-W-method (A, L, b, θ) based on the coefficients (59) and stability parameter $\theta = (3 + \sqrt{3})/6$. This method has only 2 stages, but it is of order three in ODE sense since (10) is fulfilled.

AMFR-W2 is the 2-stage AMFR-W-method (A, L, b, θ, μ) with coefficients (59), where we have chosen $\mu = \theta = (3 + \sqrt{3})/6$ for $m \leq 3$ and $\mu = 4\kappa_4\theta$, $\theta = (3 + \sqrt{3})/6$ for $m = 4$ (with $\kappa_4 = 0.2576$) to meet the stability bounds given in Theorem 9. The method is order three in ODE sense.

4.1   Linear Diffusion Equation with Constant Coefficients

In order to illustrate the stability results for AMF-type W-methods in Sect. 3 let
us consider the linear diffusion reaction partial differential equation with constant
coefficients and mixed derivative terms

$$\partial_t u = \sum_{i,j=1}^{m} \alpha_{i,j} \, \partial^2_{x_i x_j} u + g(t, \mathbf{x}), \qquad \mathbf{x} \in (0,1)^m, \quad t \in (0,1], \tag{60}$$

with $g(t, \mathbf{x})$ chosen such that

$$u(t, \mathbf{x}) = u_e(t, \mathbf{x}) := e^t \left(\prod_{j=1}^{m} x_j (1 - x_j) + \kappa \sum_{j=1}^{m} (x_j + \tfrac{1}{j+2})^2 \right) \tag{61}$$

is the exact solution of (60). The initial condition $u(0, \mathbf{x}) = u_e(0, \mathbf{x})$ and Dirichlet
boundary conditions (BCs) are imposed. We restrict our attention to the cases
$m = 3, 4$. Observe that for $\kappa = 0$ we have homogeneous boundary conditions,
but non-homogeneous time-dependent Dirichlet BCs are obtained when $\kappa = 1$.
Furthermore, we take $\alpha_{i,i} = 1$, $1 \le i \le m$, and $\alpha_{i,j} = \alpha$, for $i \ne j$, where $\alpha > 0$ is
a parameter which will be selected in order to illustrate the stability of the AMF-type
W-methods introduced in Sect. 2. In all cases, α will be chosen so that the second
order differential operator is elliptic.

We apply the MOL approach on a uniform grid with meshwidth $\Delta x_i = 1/(N +
1)$, $1 \le i \le m$, with $N = 128$ for $m = 3$ and $N = 40$ if $m = 4$. A semi-discretized
system

$$\dot{U} = \mathcal{M} U + G(t) + b(t) \tag{62}$$

of dimension N^m is obtained, where $\mathcal{M}$ is given in (13), D_{x_i} and $D_{x_i x_i}$ are
the differentiation matrices corresponding to the first and second order central
differences in each spatial direction, $G(t)$ denotes the discretization of the reaction
term $g(t, \mathbf{x})$ and $b(t)$ stores the terms due to non-homogeneous boundary conditions.
From here, the differential equation (62) also takes the form (2) and (3). Observe that
the exact solution (61) is e^t times a polynomial of degree 2 in each spatial variable
so that the global errors come only from the time discretization. Now, AMF-type
W-methods are applied to (62) with fixed stepsize $\tau = 2^{-j}$, $2 \le j \le 10$, as detailed
in Sect. 2.

The time integrations of (62) for $m = 3$ and $m = 4$ spatial dimensions with
the methods presented above are summarized in Figs. 1, 2, 3, 4, 5, and 6 below.
Figure 1 deals with the three-dimensional case, and Figs. 2 and 3 correspond to the
case $m = 4$. The global errors are plotted in relation to both the stepsize τ -to check
the temporal order of the current method- and the CPU time in seconds -to measure
the efficiency of each integrator-.

Regarding the elliptic operator in (60), we take diffusion parameters $\alpha_{i,j} = \alpha$, for $i \neq j$, with $\alpha = 0.9$. Observe that for the case of four spatial dimensions the necessary condition (50) for stability of PDE-W-methods (23) is not fulfilled. In order to meet this condition we also take $\alpha = 0.7$ when $m = 4$.

For the case $m = 3$ in Fig. 1 the methods **HV**, **MCS**, **AMFR-W1** and **AMF-W2** are seen to be second order methods as expected, whereas **PDE-W2** and **AMFR-W2** attain order three when $\kappa = 0$ (homogeneous boundary conditions). For $\kappa = 1$ the order of convergence of these two latter methods is more irregular, and it seems to be two for larger stepsizes and around 2.5 for medium and small stepsizes.

On the other hand, for the four-dimensional case, when $\alpha = 0.9$ the necessary condition (50) for stability of PDE-W-methods is not fulfilled and, in fact, **PDE-W2** is unstable in this case, see Fig. 2 (left). For the remaining methods, the selected values for the parameters θ and μ ensure stability according to Theorems 4 and 9 and the observed temporal orders of convergence are similar to those obtained when $m = 3$. When $\alpha = 0.7$ and $m = 4$, the stability requirements are satisfied for all the methods and this is illustrated in Fig. 2 (right), where a convergence order around three is observed for the methods **AMFR-W2** and **PDE-W2** in case of homogeneous boundary conditions.

In order to make a more fair comparison of the performance of the methods in a constant time-step size framework, we have plotted in Fig. 3 the global error versus the CPU time for the 4D-problem with $\alpha = 0.7$, both with homogeneous and time-dependent boundary conditions. We can observe that the **HV** is a good candidate despite being a second order method. This latter method is only outperformed by the **AMFR-W2** and **PDE-W2** methods when medium-high accuracies are required and homogeneous BCs are imposed. This could be explained by the fact that the **HV**-method has a reduced computational cost per integration step (similar to **AMFR-W1**) and it also possesses small error constants. For time dependent BCs it is possible to make a simple change in the PDE problem through multilinear interpolation (see e.g. [10]) in order to reduce the problem to homogeneous BCs, which is a more favorable situation for AMF-type W-methods.

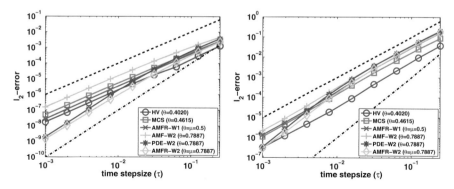

Fig. 1 3D Linear model (60)–(61) with $\alpha = 0.9$. Error vs time stepsize in the case of homogeneous boundary conditions $\kappa = 0$ (left) and in the case of time-dependent boundary conditions $\kappa = 1$ (right). $\Delta x_i = 1/129$, $1 \leq i \leq 3$

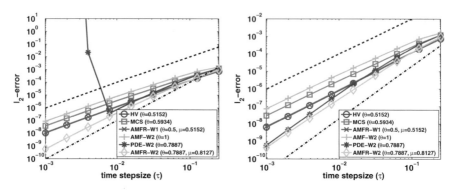

Fig. 2 4D Linear model (60) and (61) with homogeneous boundary conditions ($k = 0$) and $\Delta x_i = 1/41$, $1 \le i \le 4$: $\alpha = 0.9$ (left) and $\alpha = 0.7$ (right). Error vs time stepsize

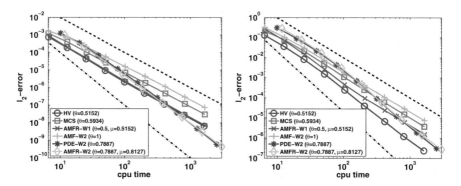

Fig. 3 4D Linear model (60)–(61) with $\alpha = 0.7$. Error vs CPU time, with homogeneous boundary conditions $k = 0$ (left) and time-dependent boundary conditions $k = 1$ (right). $\Delta x_i = 1/41$, $1 \le i \le 4$

4.2 The Heston Model

The Heston model [15] is a two-dimensional extension of the well-known Black-Scholes equation from financial option pricing theory. The results obtained in the experiments on this problem show that the proposed AMF-type schemes also perform correctly on PDEs with variable coefficients, and they can be easily applied on practical models that involve mixed derivatives terms.

This model predicts the fair price of a call option $u(s, v, t)$ at time $t > 0$, when the asset price is $s > 0$ and $v > 0$ represents its variance, by the following partial differential equation

$$
\begin{aligned}
\partial_t u = {} & \frac{1}{2} s^2 v \, \partial_{ss}^2 u + \rho \sigma s v \, \partial_{sv}^2 u + \frac{1}{2} \sigma^2 v \, \partial_{vv}^2 u \\
& + (r_d - r_f) s \, \partial_s u + \kappa (\eta - v) \, \partial_v u - r_d u.
\end{aligned}
\tag{63}
$$

Here t represents the days left until what it is called *maturity time* $T > 0$, so $t \in [0, T]$, $s > 0$, $v > 0$. The parameter $\kappa > 0$ is the mean-reversion rate and $\eta > 0$ is the long-term mean, r_d and r_f represent respectively the domestic and foreign interest rates, $\sigma > 0$ is the volatility of the variance and $\rho \in [-1, 1]$ measures the correlation between the two variables s and v.

The details of the derivation of this PDE (63) from the corresponding stochastic model can be seen in [15]. Maximum values for the spatial variables $(s, v) \in [0, S] \times [0, V]$ are prefixed and in the case of a *European call option*, the following boundary conditions are imposed

$$s = 0 : u(0, v, t) = 0, \quad t \in [0, T]$$
$$s = S : \partial_s u(S, v, t) = e^{-r_f t}, \quad t \in [0, T] \tag{64}$$
$$v = V : u(s, V, t) = se^{-r_f t}, \quad t \in [0, T]$$

On the other hand, the initial condition

$$u(s, v, 0) = \max(0, s - K) \tag{65}$$

is considered, where $K > 0$ is the *strike price* of the option, i.e., the price that the holder can buy the asset for when the option expires.

The values for the PDE parameters have been experimentally adjusted in many different practical situations. Here we will consider three different cases. The first one is the set of values proposed in [26]

$$\kappa = 0.6067, \ \eta = 0.0707, \ \sigma = 0.2928, \ \rho = -0.7571, \ r_d = 0.03, \ r_f = 0, \tag{66}$$

in such a way that the boundary conditions are time-independent. Secondly the set of values in [29]

$$\kappa = 2.5, \ \eta = 0.06, \ \sigma = 0.5, \ \rho = -0.1, \ r_d = 0.0507, \ r_f = 0.0469, \tag{67}$$

is considered, such that the boundary conditions are time-dependent with a small correlation parameter ρ. Finally we also consider a time-dependent case with a larger correlation parameter

$$\kappa = 1.5, \ \eta = 0.02, \ \sigma = 0.62, \ \rho = -0.67, \ r_d = 0.01, \ r_f = 0.02. \tag{68}$$

In the case (66) the codes perform the time integration until $T = 3$, whereas $T = 0.25$ and $T = 1$ are considered in cases (67) and (68), respectively. In all cases, we take $K = 100$, $S = 30K$ and $V = 15$.

We apply the MOL approach on this model on a non-uniform spatial mesh following the ideas given in [18], since it is known that uniform spatial grids are not efficient on it, because the initial condition (65) has a differentiability problem at $s = K$ and that for v close to 0 the PDE becomes advection-dominated. So we build a rectangular non-uniform grid

$$s_0 = 0 < s_1 < \cdots < s_m = S, \quad v_0 = 0 < v_1 < \cdots < v_n = V$$

where there are many more points close to $s = K$ and $v = 0$ than in the rest of the domain. We must take into account that, due to the boundary conditions (64), finite-differences approximations are only applied at the nodes (s_i, v_j) with $1 \leq i \leq m$ and $0 \leq j \leq n-1$. At each node of this grid, the partial derivatives of the PDE (63) are approximated by the corresponding finite-difference formulation given in detail in [18]. Roughly speaking, in the case of the derivatives $\partial_{ss}^2 u$, $\partial_{vv}^2 u$ and $\partial_s u$, second-order central differences are applied. However, due to a change in the direction of the advection for v, different formulations are used to approximate $\partial_v u$ when $0 \leq v_j \leq 1$ and $v_j > 1$. Finally, the mixed derivative $(\partial_{sv}^2 u)$ is approximated by second-order central differences for the first partial derivative at each spatial direction.

Adding the initial and boundary conditions (64) and (65) and putting all the finite differences together at each spatial point, we arrive at the following linear semi-discrete IVP of dimension $m \cdot n$ of type (2) and (3)

$$U'(t) = F(t, U) = \sum_{j=0}^{2} F_j(t, U), \quad U(0) = U_0, \quad t \in [0, T] \tag{69}$$

where

$$F_j(t, U) = A_j U + g_j e^{-r_f t}, \ (j = 1, 2), \quad F_0(t, U) = A_0 U + g_0 e^{-r_f t} - r_d U. \tag{70}$$

$F_0(t, U)$ represents the splitting term corresponding to the mixed derivatives together with the reaction part $G(U) = -r_d U$, whereas $F_j(t, U)$, $(j = 1, 2)$ corresponds to the directional splitting terms. $\{g_j\}_{j=0}^{2}$ are constant vectors that come from the time-dependent boundary conditions (64) and the constant matrices A_1 and A_2 have simple structures

$$A_1 = \text{diag}(A_1^{(0)}, A_1^{(1)}, \ldots, A_1^{(n-1)}), \quad A_2 = \tilde{A} \otimes I_m$$

where each submatrix $A_1^{(k)}$ is a tridiagonal matrix of dimension m and $\tilde{A}$ has dimension n with only five non-zero diagonals. The constant matrix A_0 of dimension $n \cdot m$ has nine non-zero bands (parallel to the main diagonal) that are non-consecutive but it is never used in the AMF (or ADI) factorizations. We mention that in [18] the splitting (69)-(70) is not applied exactly in this way, since the reaction term $G(U)$ is included in the directional terms in the following way $F_j(t, U) = A_j U + g_j e^{-r_f t} - (r_d/2)U$, $j = 1, 2$. However, this does not imply any significant change in the numerical results below in Figs. 4, 5, and 6.

Figures 4, 5, and 6 show the results for the cases (66)–(68) of the Heston problem, respectively. The time integrations have been carried out for $\tau = 2^{-j}$, $2 \leq j \leq$

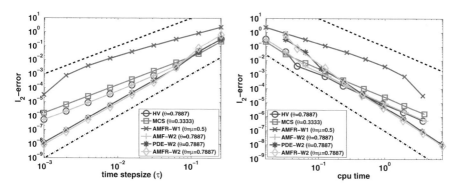

Fig. 4 Heston problem, case (66) with $m = 200$ and $n = 100$. Error vs time stepsize (left). Error vs CPU time (right)

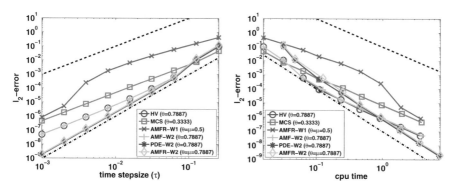

Fig. 5 Heston problem, case (67) with $m = 200$ and $n = 100$. Error vs time stepsize (left). Error vs CPU time (right)

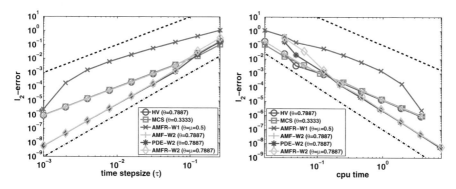

Fig. 6 Heston problem, case (68) with $m = 200$ and $n = 100$. Error vs time stepsize (left). Error vs CPU time (right)

10, and the global errors have been measured with respect to a reference solution at the respective end-point T, obtained with the DOP853 code [12] with a very stringent tolerance. In Figs. 4, 5, and 6 it is observed that all methods confirm the achieved orders on the previous constant coefficient PDE (60), i.e, order around two for **HV, MCS, AMF-W2** and order three for **AMFR-W2** and **PDE-W2**, but with the important difference that in this case the order three is maintained even in the case of time-dependent BCs. The observed order for **AMFR-W1** and larger stepsizes lies around 1.5, and order two can be observed when very small stepsizes are considered. It is also noteworthy to observe that the global errors provided by **HV** and **AMF-W2** are very similar in the three cases.

Regarding the efficiency (plots in the right side) it can be appreciated that the higher order methods, **PDE-W2** and **AMFR-W2**, are the most efficient when medium or small errors are required. For low accuracy (2 or 3 significant digits) **HV, MCS** and **AMF-W2** are better.

Acknowledgments The authors thank Ernst Hairer and Soledad Pérez-Rodríguez for the revision and their scientific contribution to the content of the current chapter, which is collected in references [9, 10].

References

1. I.J.D. Craig, A.D. Sneyd, An alternating-direction implicit scheme for parabolic equations with mixed derivatives. Comput. Math. Appl. **16**(4), 341–350 (1988)
2. J. Douglas, Jr., H.H. Rachford, Jr., On the numerical solution of heat conduction problems in two and three space variables. Trans. Am. Math. Soc. **82**, 421–439 (1956)
3. B. Düring, M. Fournié, A. Rigal, High-order ADI schemes for convection-diffusion equations with mixed derivative terms, in *Spectral and High Order Methods for Partial Differential Equations - ICOSAHOM'12*, ed. by T. Gammarth, M. Azaïez, et al. Lecture Notes in Computational Science and Engineering, vol. 95 (Springer, Berlin, 2014), pp. 217–226
4. B. Düring, J. Miles, High-order ADI scheme for option pricing in stochastic volatility models. J. Comput. Appl. Math. **316**, 109–121 (2017)
5. A. Gerisch, J.G. Verwer, Operator splitting and approximate factorization for taxis-diffusion-reaction models. Appl. Numer. Math. **42**(1–3), 159–176 (2002). Ninth Seminar on Numerical Solution of Differential and Differential-Algebraic Equations (Halle, 2000)
6. S. González Pinto, D. Hernández Abreu, S. Pérez Rodríguez, Rosenbrock-type methods with inexact AMF for the time integration of advection diffusion reaction PDEs. J. Comput. Appl. Math. **262**, 304–321 (2014)
7. S. González Pinto, D. Hernández Abreu, S. Pérez Rodríguez, R. Weiner, A family of three-stage third order AMF-W-methods for the time integration of advection diffusion reaction PDEs. Appl. Math. Comput. **274**, 565–584 (2016)
8. S. González Pinto, D. Hernández Abreu, S. Pérez Rodríguez, W-methods to stabilize standard explicit Runge-Kutta methods in the time integration of advection-diffusion-reaction PDEs. J. Comput. Appl. Math. **316**, 143–160 (2017)
9. S. González Pinto, E. Hairer, D. Hernández Abreu, S. Pérez Rodríguez, AMF-type W-methods for parabolic PDEs with mixed derivatives. SIAM J. Sci. Comput. **40**(5), A2905–A2929 (2018)
10. S. González Pinto, E. Hairer, D. Hernández Abreu, S. Pérez Rodríguez, PDE-W-methods for parabolic problems with mixed derivatives. Numer. Algorithms **78**, 957–981 (2018)

11. E. Hairer, G. Wanner, *Solving Ordinary Differential Equations II. Stiff and Differential-Algebraic Problems*. Springer Series in Computational Mathematics, vol. 14, 2nd edn. (Springer, Berlin, 1996)
12. E. Hairer, S.P. Norsett, G. Wanner, *Solving Ordinary Differential Equations I. Non Stiff Problems*. Springer Series in Computational Mathematics, vol. 8, 2nd rev. edn. (Springer, Berlin, 1993)
13. C. Hendricks, M. Ehrhardt, M. Günther, High-order ADI schemes for diffusion equations with mixed derivatives in the combination technique. Appl. Numer. Math. **101**, 36–52 (2016)
14. C. Hendricks, C. Heuer, M. Ehrhardt, M. Günther, High-order ADI finite difference schemes for parabolic equations in the combination technique with application in finance. J. Comput. Appl. Math. **316**, 175–194 (2017)
15. S.L. Heston, A closed form solution for options with stochastic volatility with applications to bonds and currency options. Rev. Financ. Stud. **6**(2), 327–343 (1993)
16. W. Hundsdorfer, Accuracy and stability of splitting with stabilizing corrections. Appl. Numer. Math. **42**(1–3), 213–233 (2002). Ninth Seminar on Numerical Solution of Differential and Differential-Algebraic Equations (Halle, 2000)
17. W. Hundsdorfer, J.G. Verwer, *Numerical Solution of Time-Dependent Advection-Diffusion-Reaction Equations*. Springer Series in Computational Mathematics, vol. 33 (Springer, Berlin, 2003)
18. K.J. in 't Hout, S. Foulon, ADI finite difference schemes for option pricing in the Heston model with correlation. Int. J. Numer. Anal. Model. **7**(2), 303–320 (2010)
19. K.J. in 't Hout, B.D. Welfert, Unconditional stability of second-order ADI schemes applied to multi-dimensional diffusion equations with mixed derivative terms. Appl. Numer. Math. **59**(3–4), 677–692 (2009)
20. K.J. in 't Hout, M. Wyns, Convergence of the modified Craig-Sneyd scheme for two-dimensional convection-diffusion equations with mixed derivative term. J. Comput. Appl. Math. **296**, 170–180 (2016)
21. T. Jax, G. Steinebach, Generalized ROW-type methods for solving semi-explicit DAEs of index-1. J. Comput. Appl. Math. **316**, 213–228 (2017)
22. J. Lang, J.G. Verwer, W-methods in optimal control. Numer. Math. **124**(2), 337–360 (2013)
23. D.W. Peaceman, H.H. Rachford, Jr., The numerical solution of parabolic and elliptic differential equations. J. Soc. Indust. Appl. Math. **3**, 28–41 (1955)
24. A. Rahunanthan, D. Stanescu, High-order W-methods. J. Comput. Appl. Math. **233**, 1798–1811 (2010)
25. J. Rang, L. Angermann, New Rosenbrock *W*-methods of order 3 for partial differential algebraic equations of index 1. BIT Numer. Math. **45**(4), 761–787 (2005)
26. W. Schoutens, E. Simons, J. Tistaert, A perfect calibration! Now what? Wilmott Mag. 66–78 (2004)
27. T. Steihaug, A. Wolfbrandt, An attempt to avoid exact Jacobian and nonlinear equations in the numerical solution of stiff differential equations. Math. Comput. **33**(146), 521–534 (1979)
28. P.J. van der Houwen, B.P. Sommeijer, Approximate factorization for time-dependent partial differential equations. J. Comput. Appl. Math. **128**(1–2), 447–466 (2001). Numerical analysis 2000, Vol. VII, Partial differential equations
29. G. Winkler, T. Apel, U. Wystup, Valuation of options in Heston's stochastic volatility model using finite element methods, in *Foreign Exchange Risk*, ed. by J. Hakala, U. Wystup (Risk Publ., London, 2002)

Two-Step W-Methods

Marcel Klinge, Helmut Podhaisky, and Rüdiger Weiner

1 Formulation of the Methods

For the numerical solution of stiff initial value problems

$$y' = f(t, y), \quad y(t_0) = y_0, \tag{1}$$

with right hand side $f : \mathbb{R} \times \mathbb{R}^n \to \mathbb{R}^n$ ROW-methods are applied frequently. There exist A- and L-stable methods. Due to their linear-implicit structure these methods are easy to implement, well-known codes are GRK4T ([2, 3]), RODAS [2] and RODASP ([2, 9]). However, due to the low stage order, order reduction can occur for very stiff problems, see e.g. [7]. Another drawback of ROW-methods is the need to compute the Jacobian in every step. W-methods allow to keep the Jacobian constant for several steps, however, because of additional order conditions, the construction of higher order methods is rather difficult.

To overcome these problems, Podhaisky et al. introduced and investigated two-step W-methods [5, 6, 10] which retain the linear-implicit structure. Recently these methods were applied in combination with approximate matrix factorization (AMF) for the solution of two-dimensional PDEs [4]. These methods can be derived from implicit two-step Runge–Kutta methods by applying one step of Newton's method.

M. Klinge · H. Podhaisky (✉) · R. Weiner
Institut für Mathematik, Martin-Luther-Universität Halle-Wittenberg, Halle, Germany
e-mail: marcel.klinge@mathematik.uni-halle.de; helmut.podhaisky@mathematik.uni-halle.de; ruediger.weiner@mathematik.uni-halle.de

© The Author(s), under exclusive license to Springer Nature Switzerland AG 2021
T. Jax et al. (eds.), *Rosenbrock—Wanner–Type Methods*, Mathematics Online
First Collections, https://doi.org/10.1007/978-3-030-76810-2_5

An s-stage two-step W-method (TSW-method) is given by

$$Y_{m,i} = u_m + h_m \sum_{j=1}^{s} a_{ij} k_{m-1,j} + h_m \sum_{j=1}^{i-1} \tilde{a}_{ij} k_{m,j},$$

$$(I - h_m \gamma T_m) k_{m,i} = f(t_{m,i}, Y_{m,i}) + h_m T_m \sum_{j=1}^{s} \gamma_{ij} k_{m-1,j} + h_m T_m \sum_{j=1}^{i-1} \tilde{\gamma}_{ij} k_{m,j},$$

$$u_{m+1} = u_m + h_m \sum_{j=1}^{s} (b_j k_{m,j} + v_j k_{m-1,j}).$$

$$(2)$$

Here u_m is an approximation to the exact solution $y(t_m)$. The matrix T_m is arbitrary. The order of the method is independent of the choice of T_m. For stability reasons it should be an approximation to the Jacobian $f_y(t_m, u_m)$, but it can be kept constant for some steps in practical computations. The s stage values $Y_{m,i}$ are approximations to $y(t_{m,i})$, and $k_{m,i}$ represent approximations to the corresponding stage derivatives. We always assume that the nodes c_i are pairwise distinct with $t_{m,i} = t_m + c_i h_m$ and $c_s = 1$. The coefficients $a_{ij}, \tilde{a}_{ij}, \gamma_{ij}$ and $\tilde{\gamma}_{ij}$ with $\tilde{\gamma}_{ii} = \gamma > 0$ can be collected in matrices $A = (a_{ij})_{i,j=1}^{s}, \tilde{A} = (\tilde{a}_{ij}), \Gamma = (\gamma_{ij}), \tilde{\Gamma} = (\tilde{\gamma}_{ij})$ and vectors $b = (b_i)_{i=1}^{s}, v = (v_i), c = (c_i)$. $\tilde{A}$ and $\tilde{\Gamma}$ are strictly lower triangular matrices. Note that some of the coefficients will depend on the step size ratio $\sigma_m = h_m / h_{m-1}$. We will omit the index m for a shorter notation, i.e., $\sigma = \sigma_m$ and $A = A_m$ etc.

The methods are linearly implicit. For every stage $i = 1, \ldots, s$ a system of linear equations has to be solved with a coefficient matrix which is constant within the step, i.e., only one LU decomposition is required per step.

Due to the two-step character the methods require additional starting values $k_{0,i}$, $i = 1, \ldots, s$.

2   Order Conditions and Stability

Here we collect some results from [6] and [5]. Order conditions can be derived by inserting the exact solution and studying the Taylor series expansions of the residuals. The residual errors can be analysed with the help of the following simplifying assumptions:

$$C(q): \qquad \frac{\sigma^l c_i^l}{l!} = \sigma \sum_{j=1}^{s} \frac{a_{ij}(c_j - 1)^{l-1}}{(l-1)!} + \sigma^l \sum_{j=1}^{i-1} \frac{\tilde{a}_{ij} c_j^{l-1}}{(l-1)!}, \qquad (3)$$

$$l = 1, \ldots, q, \quad i = 1, \ldots, s,$$

$$\Gamma(q): \quad -\frac{\gamma \sigma^l c_i^{l-1}}{(l-1)!} = \sigma \sum_{j=1}^{s} \frac{\gamma_{ij}(c_j-1)^{l-1}}{(l-1)!} + \sigma^l \sum_{j=1}^{i-1} \frac{\widetilde{\gamma}_{ij} c_j^{l-1}}{(l-1)!}, \quad (4)$$

$$l = 1, \ldots, q, \quad i = 1, \ldots, s,$$

$$B(p): \quad \frac{\sigma^l}{l!} = \sigma^l \sum_{i=1}^{s} \frac{b_i c_i^{l-1}}{(l-1)!} + \sigma \sum_{i=1}^{s} \frac{v_i(c_i-1)^{l-1}}{(l-1)!}, \quad (5)$$

$$l = 1, \ldots, p.$$

The method is said to be of stage order q if $C(q)$ and $\Gamma(q)$ are satisfied. We denote the errors of the starting values by

$$\varepsilon_0 = \|y(t_0) - u_0\|, \quad v_0 = \max_{i=1,\ldots,s} \|y'(t_0 + c_i h_0) - k_{0,i}\|.$$

Two-step W-methods are stable for $h \to 0$ by design and hence convergence follows without additional stability conditions. Analogously to [6] one shows the following theorem.

Theorem 1 *Assume that the initial errors satisfy $\varepsilon_0 = \mathcal{O}(h_0^p)$ and $v_0 = \mathcal{O}(h_0^q)$ with $p, q \in \mathbb{N}$. Let the coefficients of the method and the step size ratio be bounded, i.e. $\sigma_m = h_m/h_{m-1} < \sigma_{\max}$. If the method (2) satisfies the simplifying assumptions $C(q)$, $\Gamma(q)$ and $B(p)$, then for arbitrary matrices T_m it is convergent of order $p^* = \min(q+1, p)$ for a sufficiently smooth right hand side f.*

The simplifying conditions (3)–(5) are linear relations between the coefficient matrices. The Vandermonde–type matrices constructed from the nodes are quadratic (and non-singular) in the convenient case $p = q = s$. Then we can satisfy the order conditions by solving for A, Γ and $v^\top$:

$$A = (CV_0 D^{-1} - \widetilde{A} V_0) S V_1^{-1} \quad (6)$$

$$\Gamma = -(\gamma I + \widetilde{\Gamma}) V_0 S V_1^{-1} \quad (7)$$

$$v^\top = (\mathbb{1}^\top D^{-1} - b^\top V_0) S V_1^{-1}. \quad (8)$$

Here we used the notation

$$V_0 = (c_i^{j-1}), \quad V_1 = ((c_i-1)^{j-1}), \quad D = \text{diag}(i), \quad S = \text{diag}(\sigma^{i-1}), \quad C = \text{diag}(c_i).$$

The system (6)–(8) is also used in the implementation to re-compute the coefficients when the step size changes.

To study stability, we apply the method (2) to the usual test equation $y' = \lambda y$ with $T_m = \lambda$ and obtain the matrix recursion

$$\begin{pmatrix} h_m K_m \\ u_{m+1} \end{pmatrix} = M(z) \begin{pmatrix} h_{m-1} K_{m-1} \\ u_m \end{pmatrix} \tag{9}$$

where the $(s + 1) \times (s + 1)$ amplification matrix $M(z)$ is given by

$$M(z) = \begin{pmatrix} \sigma W(z)\beta & W(z)\mathbb{1} \\ \sigma (b^\top W(z)\beta + v^\top) & 1 + b^\top W(z)\mathbb{1} \end{pmatrix} \tag{10}$$

with $K_m = [k_{m1}, \ldots, k_{ms}]^\top$, $\beta := A + \Gamma$, $\tilde{\beta} := \tilde{A} + \tilde{\Gamma}$ and $W(z) = [(1 - z\gamma)I - z\tilde{\beta}]^{-1}z$.

We now consider constant step sizes. Then stability is characterized by the spectral radius of M.

Definition 1 We call the set $\mathscr{S} = \{z \in \mathbb{C} : \rho(M(z)) < 1\}$ stability domain of the TSW-method. The method is called A(α)-stable if $\{z \in \mathbb{C} : |\arg(z) - \pi| \le \alpha\} \subseteq \bar{\mathscr{S}}$. It is said to be A-stable if $\alpha = \frac{\pi}{2}$. We call a method *stiffly accurate* if for all fixed $u_m, k_{m-1,i}, \quad i = 1, \ldots, s$, the condition

$$\lim_{|z| \to \infty} u_{m+1} = 0 \tag{11}$$

holds. A method is called L(α)-stable if it is A(α)-stable and (11) is fulfilled. It is said to be L-stable if $\alpha = \frac{\pi}{2}$.

Stiff accuracy is equivalent by a vanishing last row in the stability matrix $M(z)$ for $|z| \to \infty$. This can also be achieved for variable step sizes. It is proved in [5] that a TSW-method (2) with $c_s = 1$ which satisfies the simplifying conditions $C(s)$, $\Gamma(s)$, $B(s)$ is stiffly accurate if and only if

$$b^\top = e_s^\top (\gamma I + \tilde{A} + \tilde{\Gamma}) \tag{12}$$

$$v^\top = e_s^\top (A + \Gamma) \tag{13}$$

holds, where e_s denotes the s-th unit vector.

In [5] stiffly accurate methods of order $p^* = s$ with $\tilde{\Gamma} = 0$ were constructed. However, Theorem 1 shows that the method will have order of convergence $p^* = s + 1$ if we can satisfy $B(s + 1)$ for variable σ. In the following section we will give additional conditions which allow to construct stiffly accurate TSW-methods which satisfy $B(s + 1)$ and thus have convergence order $p^* = s + 1$ for variable step sizes.

3   Stiffly Accurate Methods of Order $p^* = s + 1$

Assume that (6)–(8) are satisfied, i.e. we have a method of order $p^* = s$. By (8) the simplifying assumption $B(s)$ is fulfilled. It remains to satisfy

$$\frac{\sigma^{s+1}}{(s+1)!} = \frac{\sigma^{s+1}}{s!} b^\top c^s + \frac{\sigma}{s!} v^\top (c - 1)^s,$$

where $c^s = (c_1^s, \ldots, c_s^s)^\top$. Substituting (8) this leads to

$$\sigma^s = (s+1)\sigma^s b^\top c^s + (s+1)(1^\top D^{-1} - b^\top V_0) S V_1^{-1} (c - 1)^s. \tag{14}$$

Due to the appearance of S this condition must be satisfied for variable σ. This leads to $s + 1$ conditions for the coefficients at powers σ^l for $l = 0, \ldots, s$. It turns out that it is possible to find stiffly accurate methods which in addition to (6)–(8) also satisfy (14) for all σ.

Theorem 2 *Let the TSW-method satisfy the conditions $\Gamma(s)$, $C(s)$ and $B(s)$ and let $c_s = 1$. Then under the conditions*

$$b^\top = \left(\frac{1}{2}, \ldots, \frac{1}{s+1}\right) V_0^{-1} C^{-1} \tag{15}$$

and

$$(\tilde{\gamma}_{s,1}, \ldots, \tilde{\gamma}_{s,s-1}, \gamma) = \left(\frac{1}{2}, \ldots, \frac{1}{s+1}\right) V_0^{-1} C^{-1} - e_s^\top \tilde{A} \tag{16}$$

the method satisfies $B(s + 1)$ and is stiffly accurate.

Proof For $B(s + 1)$ we have to show that (14) is satisfied. We denote $x = V_1^{-1}(c - 1)^s$. Considering powers of σ condition (14) is equivalent to the $s + 1$ conditions

$$0 = \left(\frac{1}{l} - b^\top c^{l-1}\right) x_l, \quad l = 1, \ldots, s$$

$$0 = \frac{1}{s+1} - b^\top c^s. \tag{17}$$

For $l = 1$ we have

$$x_1 = e_1^\top x = e_1^\top V_1^{-1}(c - 1)^s = e_s^\top (c - 1)^s = 0$$

by the definition of V_1 and because of $c_s = 1$. It is

$$C V_0 = (c, c^2, \ldots, c^s)$$

and

$$b^\top C V_0 = (b^\top c, \ldots, b^\top c^s).$$

On the other hand with (15) it holds

$$b^\top C V_0 = (\frac{1}{2}, \ldots, \frac{1}{s+1}).$$

Consequently, the remaining s conditions of (17) for $l = 2, \ldots, s+1$ are fulfilled, i.e. $B(s+1)$ holds.

To prove stiff accuracy we have to show that (12) and (13) are satisfied. It holds

$$e_s^\top (\widetilde{A} + \gamma I + \widetilde{\Gamma}) = e_s^\top \widetilde{A} + (\widetilde{\gamma}_{s,1}, \ldots, \widetilde{\gamma}_{s,s-1}, \gamma)$$

$$= e_s^\top \widetilde{A} + (\frac{1}{2}, \ldots, \frac{1}{s+1}) V_0^{-1} C^{-1} - e_s^\top \widetilde{A} \quad \text{by (16)}$$

$$= b^\top \text{ by (15)},$$

i.e. (12) is fulfilled.

With (6) and (7) we have

$$e_s^\top (A + \Gamma) = e_s^\top (C V_0 D^{-1} - \widetilde{A} V_0 - \gamma V_0 - \widetilde{\Gamma} V_0) S V_1^{-1}$$

$$= (\mathbb{1}^\top D^{-1} - b^\top V_0) S V_1^{-1} \text{ (by } e_s^\top C V_0 = e_s^\top V_0 = \mathbb{1}^\top \text{ and (12))}$$

$$= v^\top \quad \text{by (8)},$$

i.e. (13) is satisfied, and thus the method is stiffly accurate.

Corollary 1 *A TSW–method which satisfies the assumptions of Theorem 2 is convergent of order $p^* = s + 1$ for variable step sizes if the initial errors satisfy $\varepsilon_0 = \mathcal{O}(h_0^{s+1})$, $v_0 = \mathcal{O}(h_0^s)$, and the step size ratio is bounded, i.e. $\sigma_m = h_m / h_{m-1} < \sigma_{\max}$.*

Proof Theorem 2 guarantees $B(s+1)$ and thus Theorem 1 can be applied.

Remark 1 By (15) b is uniquely defined by the nodes c_i and is independent of the step size ratio σ.

Despite the additional conditions (15) and (16) there are free parameters left, namely $c_1, \ldots, c_{s-1}$, $\widetilde{A}$ and $\widetilde{\Gamma}$ (except the last row). In the next section we will use these parameters to construct stiffly accurate methods of order $s + 1$ with good stability properties.

4   Construction of Methods with $B(s + 1)$

In this section we describe the construction of TSW-methods, which satisfy the
conditions (6)–(8), (15) and (16), i.e. the methods have the properties

$$C(s), \ \Gamma(s), \ B(s + 1), \ c_s = 1, \ \text{stiffly accurate.}$$

The free parameters are $\widetilde{A}$, $\widetilde{\gamma}_{ij}$, $i = 1, \ldots, s-1$, $j = 1, \ldots, i-1$ and $c_1, \ldots, c_{s-1}$.
We search for suitable methods with large angle α of L(α)-stability, small error
constants and small spectral radius at infinity $\varrho(M(\infty))$ for $\sigma = 1$. For stiffly
accurate TSW-methods this is equivalent to $\varrho(G_\infty)$, where

$$G_\infty := W_\infty \beta, \quad \text{with } W_\infty := W(\infty) = -(\gamma I + \widetilde{\beta})^{-1}.$$

We consider as error constant *ferr* the sum of the magnitude of the residual errors
in $C(s + 1)$, $\Gamma(s + 1)$ and $B(s + 2)$. It is also our aim to have small coefficients of
the methods. The optimization is carried out for constant step sizes, i.e. for $\sigma = 1$
and we use fmincon from the optimization toolbox in MATLAB and for algebraic
computations the computer algebra system MAXIMA. In the following we discuss
the construction of TSW–methods for different numbers of stages s in detail.

4.1   Methods with $s = 2$ Stages

In this case we have only two free parameters, namely, c_1 and $\widetilde{a}_{21}$. One can
determine c_1 with respect to the angle α and the spectral radius $\varrho(G_\infty)$ for $\sigma = 1$.
Figure 1 shows the angle α and $\varrho(G_\infty)$ as a function of the node c_1. Note, that one
can find methods with $\alpha = 90°$ only for $c_1 > 1$. With fixed c_1 we determine $\widetilde{a}_{21}$ of

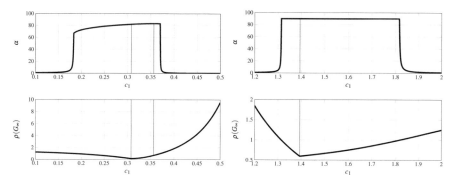

Fig. 1 Angle α of L(α)-stability vs. c_1 (*top*) and $\varrho(G_\infty)$ vs. c_1 (*bottom*), with $c_1 < 1$ (*left*) and
$c_1 > 1$ (*right*)

corresponding methods to have small error constants and coefficients. We present the following two methods:

2a:

$$c_1 = 3.078214324506323200e-1, \quad c_2 = 1, \qquad \tilde{a}_{21} = 2.069078866037454400e+0,$$

$$\tilde{\gamma}_{21} = -1.286853766869382900e+0, \quad \gamma = 2.592143494752462400e-1.$$

2b:

$$c_1 = 3.445020153831068200e-1, \quad c_2 = 1, \qquad \tilde{a}_{21} = 1.766481521486239500e+0,$$

$$\tilde{\gamma}_{21} = -1.028431797882353400e+0, \quad \gamma = 2.457403827655164100e-1.$$

4.2 Methods with $s = 3$ Stages

In the case of three stages we have six free parameters: $\tilde{a}_{21}, \tilde{a}_{31}, \tilde{a}_{32}, \tilde{\gamma}_{21}, c_1$ and c_2. Analogous to [5] we want to have methods with $\varrho(G_\infty) = 0$ for $\sigma = 1$. The characteristic polynomial is given by $\det(xI - G_\infty) = x^3 + p_2 x^2 + p_1 x + p_0$. We compute $\tilde{a}_{21}, \tilde{a}_{31}$ and $\tilde{a}_{32}$ so that $p_2 = p_1 = p_0 = 0$ are satisfied. Note that $p_2 = p_1 = 0$ is a linear system in $\tilde{a}_{31}$ and $\tilde{a}_{32}$. Inserting this into $p_0 = 0$ yields a condition for $\tilde{a}_{21}$. These calculations are done with the computer algebra system MAXIMA. Note, that we have to add a further constraint in the optimization process to satisfy condition (16). Then we optimize $\tilde{\gamma}_{21}, c_1$ and c_2 with respect to the angle α of L(α)-stability, error constant and the magnitude of the coefficients, whereby we apply different heuristics to find appropriate methods. The following method 3b has been found in this way:

3b:

$$c_1 = 4.2451803798618165e-1, \quad c_2 = 1.2555618550820942e+0, \quad c_3 = 1,$$

$$\tilde{a}_{21} = 5.1774789773658938e+0, \quad \tilde{a}_{31} = 6.3391015556851371e-1, \quad \tilde{a}_{32} = -4.0773189037882983e-2,$$

$$\tilde{\gamma}_{21} = -4.3034644907058750e+0, \quad \tilde{\gamma}_{31} = -1.3659849627611041e-2, \quad \tilde{\gamma}_{32} = -6.4041956977805674e-3,$$

$$\gamma = 2.9592668175830239e-1.$$

Additionally, we present a method, for which $\varrho(G_\infty) = 0$ does not hold, but with larger angle α than 3b, see Table 4. This method is obtained by numerical search with fmincon and with respect to the described properties above. The coefficients of the method are given by

3a:

$$c_1 = 2.7585435173749423e-1, \quad c_2 = 1.2974145641639010e+0, \quad c_3 = 1,$$

$$\tilde{a}_{21} = 4.6146103121913240e-1, \quad \tilde{a}_{31} = -6.3013501027799779e-1, \quad \tilde{a}_{32} = 3.3481277271620247e-1,$$

$$\tilde{\gamma}_{21} = 1.0038467404049227e+0, \quad \tilde{\gamma}_{31} = 1.2814081673484539e+0, \quad \tilde{\gamma}_{32} = -4.2958347323894375e-1,$$

$$\gamma = 4.4330035256651801e-1.$$

4.3 Methods with s = 4 and s = 5 Stages

In the case of four and five stages we perform the numerical optimization with different strategies to find good parameter sets. Here we have no explicit conditions for $\varrho(G_\infty) = 0$, but we want to satisfy $\varrho(G(\infty)) < 1$ for $\sigma = 1$. Again, we optimize concerning the angle α for stability, error constants and coefficients. The coefficients of the methods are shown in Tables 1, 2 and 3.

Table 1 Coefficients of method 4a

$c_1 = 3.4475069518575380e-1,$ $c_2 = -3.0199601869781884e-1,$ $c_3 = 1.2715954631040773e+0,$

$c_4 = 1,$

$\tilde{a}_{21} = -1.3807276352109585e-1,$ $\tilde{a}_{31} = 4.0288429533730259e+0,$ $\tilde{a}_{32} = -1.6608358550657365e+0,$

$\tilde{a}_{41} = 5.5395665635891145e-1,$ $\tilde{a}_{42} = 5.7259556650406740e-1,$ $\tilde{a}_{43} = 1.7058748218129905e-2,$

$\tilde{\gamma}_{21} = -1.3109542641248575e-1,$ $\tilde{\gamma}_{31} = -2.7740318778345143e+0,$ $\tilde{\gamma}_{32} = 1.1944608079043511e+0,$

$\tilde{\gamma}_{41} = 1.4615607370092432e-1,$ $\tilde{\gamma}_{42} = -5.4352839808888898e-1,$ $\tilde{\gamma}_{43} = -7.4801424301146488e-2,$

$\gamma = 3.4083914367433077e-1.$

Table 2 Coefficients of method 4b

$c_1 = 2.4902046482054652e-1,$ $c_2 = 1.8463585014782384e+0,$ $c_3 = 1.2904402196609168e+0,$

$c_4 = 1,$

$\tilde{a}_{21} = 1.2369099563404959e+0,$ $\tilde{a}_{31} = 4.6203540002585880e-1,$ $\tilde{a}_{32} = -9.1462206621367961e-2,$

$\tilde{a}_{41} = -2.7636893446018787e-2,$ $\tilde{a}_{42} = -1.6369452680547052e-2,$ $\tilde{a}_{43} = -6.4152678919227064e-3,$

$\tilde{\gamma}_{21} = 1.2850995505590568e+0,$ $\tilde{\gamma}_{31} = 5.3577018410535193e-1,$ $\tilde{\gamma}_{32} = -3.9108197137041377e-3,$

$\tilde{\gamma}_{41} = 6.2457914347561516e-1,$ $\tilde{\gamma}_{42} = 3.4191540363782635e-2,$ $\tilde{\gamma}_{43} = -2.1472697867924981e-1,$

$\gamma = 6.0381404956018603e-1.$

Table 3 Coefficients of method 5a

$c_1 = 3.246587185388872300e-1,$ $c_2 = -5.720591706090348800e-1,$ $c_3 = -1.109921351135201300e-1,$

$c_4 = 1.300474300552631400e+0,$ $c_5 = 1,$

$\tilde{a}_{21} = 5.974835146040646800e-1,$ $\tilde{a}_{31} = 8.490019260372140600e-2,$ $\tilde{a}_{32} = 5.309451223111111300e-1,$

$\tilde{a}_{41} = 8.882787859501643000e-1,$ $\tilde{a}_{42} = 4.914790217702752500e-1,$ $\tilde{a}_{43} = 1.267927289475134800e-2,$

$\tilde{a}_{51} = 5.615346901779065800e-1,$ $\tilde{a}_{52} = 6.297421387214541300e-1,$ $\tilde{a}_{53} = -6.189311019415895100e-1,$

$\tilde{a}_{54} = -1.341191447532984700e-1,$

$\tilde{\gamma}_{21} = -1.428149318299409800e-1,$ $\tilde{\gamma}_{31} = -1.387781348022771900e-1,$ $\tilde{\gamma}_{32} = -5.703644076283118600e-1,$

$\tilde{\gamma}_{41} = 1.063509214355987900e+0,$ $\tilde{\gamma}_{42} = -3.033042031892074200e-1,$ $\tilde{\gamma}_{43} = 7.049260816587147300e-1,$

$\tilde{\gamma}_{51} = 3.960037509580768300e-1,$ $\tilde{\gamma}_{52} = -6.504398625148823900e-1,$ $\tilde{\gamma}_{53} = 1.229735679813108700e+0,$

$\tilde{\gamma}_{54} = 9.975876229422198100e-2,$

$\gamma = 2.897657726225649800e-1.$

Table 4 Some properties of TSW-methods: stages s, order p, angle α of L(α)-stability, spectral radius $\varrho(G_\infty)$, error constant *ferr* and maximal magnitude of the coefficients, new methods (*top*) and TSW-methods from [5] (*bottom*)

Name	s	p	α	$\varrho(G_\infty)$	*ferr*	*maxcoeff*
2a	2	3	81.85	0.1699	0.3354	2.0690
2b	2	3	83.00	0.4907	0.3459	1.7664
3a	3	4	88.68	0.1746	4.6259	4.7382
3b	3	4	76.81	0.0000	1.7578	5.3985
4a	4	5	86.09	0.4832	10.8643	4.8077
4b	4	5	89.87	0.4690	15.6969	16.0839
5a	5	6	74.27	0.5842	33.2437	12.4194
TSW2B	2	3	82.75	0.3333	1.7778	1.0000
TSW3A	3	3	90.00	0.0000	5.3590	3.2000
TSW3B	3	3	83.49	0.0000	2.7344	3.5000

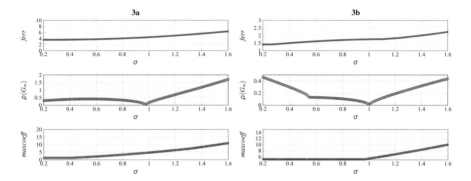

Fig. 2 Error constant *ferr* (*top*), spectral radius $\varrho(G_\infty)$ (*middle*), maximal magnitude of the coefficients *maxcoeff* (*bottom*) vs. step size ratio σ, for method 3a (*left*) and 3b (*right*)

4.4   Properties of the Methods

Properties of the optimized methods are listed in Table 4. We include for comparison the TSW-methods `TSW2B`, `TSW3A` and `TSW3B` from [5].

The numerical search is done for constant step sizes, i.e. $\sigma = 1$, but the characteristics of the methods under step sizes changes are important, too. Therefore we illustrate for 3a and 3b the error constant *ferr*, $\varrho(G_\infty)$ and the maximal magnitude of the coefficients *maxcoeff* as function of σ with $0.2 \leq \sigma \leq 1.6$, which is the crucial range for practical computations, cf. Fig. 2.

5   Numerical Tests

In this section we test the methods in MATLAB. We compare them with the two-step W-methods from [5] and with `ode23s`. The code `ode23s` from the MATLAB ODE-suite [8] is based on a Rosenbrock formula of order two and uses a third order method for error estimation and step size control. Furthermore, for comparison, we have implemented the ROW-method RODAS in MATLAB. RODAS is an L-stable ROW-method of order four with an embedded method of order three [2].

First, we illustrate the effect of order reduction in ROW-methods. We consider the van der Pol equation

$$y' = z$$
$$\varepsilon z' = (1 - y^2)z - y, \quad \varepsilon = 10^{-5}, \quad 0 \le t \le 0.5$$
$$y(0) = 2, \quad z(0) = 0.$$

Figure 3 shows the error at the endpoints vs. the constant step size h. The order reduction of RODAS is clearly observed as it behaves like second order numerically. The TSW-methods, however, are not affected at all which can be explained with their high stage order.

For the tests with step size control instead of (2) a transformed formulation

$$(I - h_m \gamma T_m)(k_{m,i} + \xi_{m,i}) = f(t_{m,i}, Y_{m,i}) + \xi_{m,i}, \tag{18}$$

where $\xi_{m,i} = \frac{1}{\gamma} \left(\sum_{j=1}^{s} \gamma_{ij} k_{m-1,j} + \sum_{j=1}^{i-1} \tilde{\gamma}_{ij} k_{m,j} \right)$ is used which avoids matrix-vector multiplications. In all cases considered we use LU decomposition for the

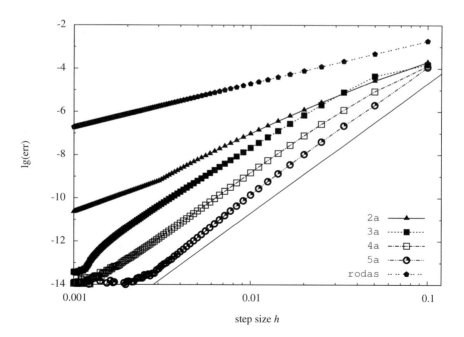

Fig. 3 Illustration of order reduction

solution of the linear systems. The Jacobian $f_y(t_m, u_m)$ is computed numerically in each step. The starting values are computed with RODAS. For error estimation we follow [5]. We calculate the weights of an embedded method by

$$b_e^\top := 0.5b^\top, \quad v_e^\top := \left(\left(\mathbb{1}^\top + 0.2e_s^\top\right)SD^{-1} - b_e^\top V_0 S\right)V_1^{-1} \tag{19}$$

and obtain an embedded solution

$$\tilde{u}_{m+1} = u_m + h_m \sum_{j=1}^{s}(b_{e,j}k_{m,j} + v_{e,j}k_{m-1,j}). \tag{20}$$

Because of the term $0.2e_s^\top$ in Eq. (19) for v_e, condition $B(s)$ is not satisfied, i.e. the embedded method is of order $p_e^* = s - 1$. Then the local error err_{emb} is estimated by

$$err_{emb} = \max_{i=1,\dots,n} \frac{\left|u_{m+1,i} - \tilde{u}_{m+1,i}\right|}{atol + rtol\left|u_{m,i}\right|}, \tag{21}$$

where $atol$ and $rtol$ denote the absolute and relative tolerances. In our tests we used the tolerances $atol = rtol = 10^{-2}, \dots, 10^{-10}$. For ode23s and the two-stage TSW-methods for some problems the tolerance 10^{-10} is omitted because of high computing time.

We used the following test problems which can be found in [2]:

- HIRES, a stiff system of eight nonlinear ordinary differential equations with $t_e = 321.8122$,
- PLATE, a linear and nonautonomous system of differential equations of dimension $n = 80$ with $t_e = 7$,
- OREGO, a stiff system of three nonlinear ordinary differential equations with $t_e = 360$,
- VDPOL, the van der Pol oscillator of two ordinary differential equations with $\varepsilon = 10^{-6}$ and two different endpoints $t_e = 2$ and $t_e = 11$.

Furthermore, we used a higher dimensional problem, i.e. the semidiscretized Burgers equation (BURGERS) from [1]:

$$u_t = vu_{xx} + uu_x + \varphi(t, x), \quad -1 \le x \le 1, \quad 0 \le t \le 2$$

$$u(0, x) = \sin(\pi(x + 1)), \quad \text{with homogeneous Dirichlet BC}$$

$$\varphi(t, x) = r(x) * \sin(t), \quad r(x) = \begin{cases} 0, & -1 \le x \le -1/3 \\ 3(x + 1/3), & -1/3 \le x \le 0 \\ 3(2/3 - x)/2, & 0 \le x \le 2/3 \\ 0, & 2/3 \le x \le 1 \end{cases}$$

with $\nu = 0.1$. The central differences of second order with $\Delta x = 1/2500$ is used for the spatial discretization. For the TSW-methods we used the discretization of the diffusion part as constant Jacobian. RODAS needs the exact Jacobian in each time step. We computed the Jacobian using Numjac were we exploit the band structure by providing JPattern.

Reference solutions for all problems are computed with high accuracy with ode15s. The errors are computed at the endpoint t_e in a weighted maximum norm

$$err = \max_{i=1,\ldots,n} \frac{|u_{m+1,i} - y_{\text{ref},i}|}{1 + |y_{\text{ref},i}|} \tag{22}$$

In Figs. 4, 5, 6, 7, 8 and 9 we present the computation time vs. the logarithm of the errors at the endpoint. The results show the potential of the TSW-methods. The methods of order three are clearly superior to ode23s. RODAS is the best method for crude tolerances, where it requires a small number of steps. Furthermore, RODAS can quickly adjust the step size to the dynamics of the solution as the step size can be enlarged by a factor 5 in a single step. For the TSW-method we need to be more conservative and restrict this factor to 1.5 (for the 5-stage method to 1.1) to avoid large changes in the coefficients A and $\tilde{A}$. For more stringent tolerances,

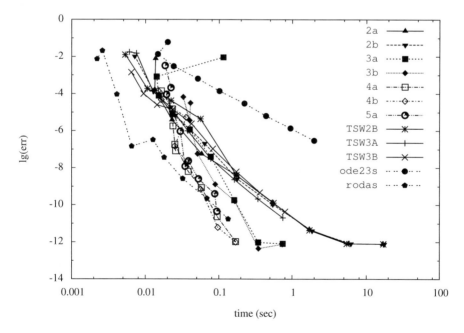

Fig. 4 Results for HIRES

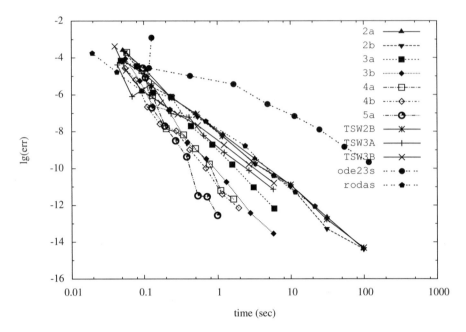

Fig. 5 Results for PLATE

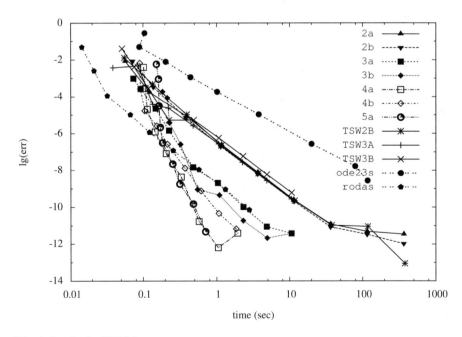

Fig. 6 Results for OREGO

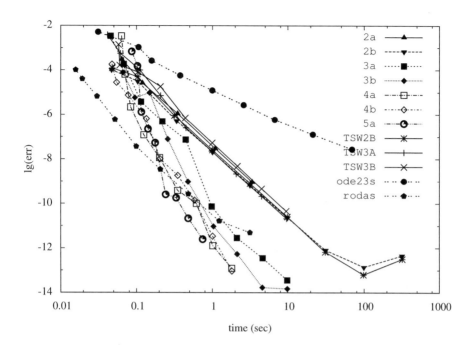

Fig. 7 Results for VDPOL with $t_e = 2$

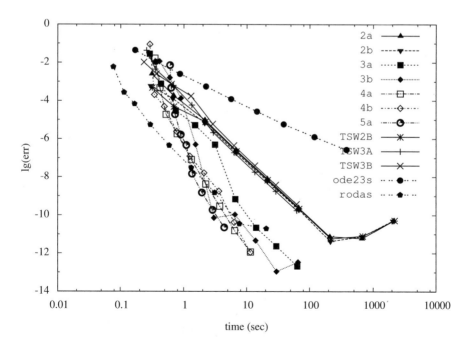

Fig. 8 Results for VDPOL with $t_e = 11$

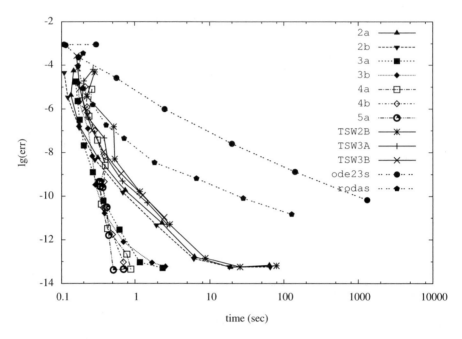

Fig. 9 Results for BURGERS

however, the higher order TSW-methods become superior, especially for PLATE. For the higher dimensional problem BURGERS the possibility to use a constant Jacobian for the TSW-methods clearly pays off.

In theses tests, we computed for all methods the Jacobian at every step excepted for BURGERS for the TSW-methods. This is necessary for RODAS and ode23s, but not for the TSW-methods. Their order is independent of the choice of the matrix T_m. Using the same Jacobian for several consecutive steps may reduce the computing time. To illustrate this potential advantage Fig. 10 shows the results for VDPOL with $t_e = 11$ when the Jacobian is computed each step (2b,3b,4b) and when it is computed only every second step (2b-2,3b-2,4b-2). We note that a more sophisticated strategy for the recomputation of the Jacobian requires further investigations.

6   Conclusions

We have reviewed the construction of two-step W-methods. Using a new condition to satisfy $B(s + 1)$ a family of methods which are convergent of order $p = s + 1$ is derived. Furthermore, these methods can be stiffly accurate and $L(\alpha)$-stable. Methods with 2–5 stages have been constructed. Numerical tests show that the new two-step W-methods are an efficient alternative to traditional ROW-methods.

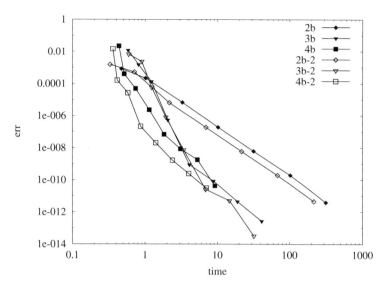

Fig. 10 Results for VDPOL with $t_e = 11$. For methods with suffix '-2' the Jacobian was computed only every second step

Rosenbrock-type peer methods introduced in [7] have some similarities (high stage order, two-step scheme) with TSW-method but neither class can, in general, be rewritten in terms of the other.

References

1. M.P. Calvo, J. de Frutos, J. Novo, Linearly implicit Runge–Kutta methods for advection–reaction–diffusion equations. Appl. Numer. Appl. **37**, 535–549 (2001)
2. E. Hairer, G. Wanner, *Solving Ordinary Differential Equations II. Stiff and Differential–Algebraic Problems* (Springer, Berlin, 1996)
3. P. Kaps, P. Rentrop, Generalized Runge–Kutta methods of order four with step size control for stiff ordinary differential equations. Numer. Math. **38**, 55–68 (1979)
4. M. Klinge, D. Hernández-Abreu, R. Weiner, A comparison of one–step and two–step methods with approximate matrix factorization. J. Comput. Appl. Math. **387**, 112519 (2021). https://doi.org/10.1016/j.cam.2019.112519
5. H. Podhaisky, R. Weiner, B.A. Schmitt, Two-step W-methods for stiff ODE systems. Vietnam J. Math. **30**, 591–603 (2002)
6. H. Podhaisky, B.A. Schmitt, R. Weiner, Design, analysis and testing of some parallel two-step W-methods for stiff systems. Appl. Numer. Math. **42**, 381–395 (2002)
7. H. Podhaisky, R. Weiner, B.A. Schmitt, Rosenbrock–type 'Peer' two-step methods. Appl. Numer. Math. **53**, 409–420 (2005)
8. L.F. Shampine, M.W. Reichelt, The MATLAB ODE suite. SIAM J. Sci. Comput. **18**(1), 1–22 (1997)

9. G. Steinebach, P. Rentrop, An adaptive method of lines approach for modelling flow and transport in rivers, in *Adaptive Method of Lines*, ed. by A.Vande Wouver, Ph. Sauces, W.E. Schiesser (Chapman & Hall/CRC, Boca Raton, 2001), pp. 181–205
10. R. Weiner, B.A. Schmitt, H. Podhaisky, *Parallel Two-Step W-Methods on Singular Perturbation Problems*. Lecture Notes in Computer Science, vol. 2328 (Springer, Berlin, 2002), pp. 778–785

Printed in the United States
by Baker & Taylor Publisher Services